THE LAST OF THE HUNTERS

PETER MORTIMER

First published 1987 by North Tyneside Libraries & Arts Dept., Central Library, Northumberland Square, North Shields, Tyne & Wear. Tel: Tyneside (091) 258 2811

Printed by E.F. Peterson & Son, Laygate, South Shields, Tyne & Wear

ISBN 0906 529042

Book cover and design by Helen Sinden

Fishing boat logo by Nicola Balfour

Front Cover photograph of Ollie Allcock and Mick Nugent on the Castle Dawn

supported by NORTHERN ARTS

NUCLEAR-FREE ZONE North TYNESIDE Metropolitan Borough Council

CONTENTS

ACKNOWLEDGEMENTS

Many people and organisations helped towards making this book a reality. Lucy Milton and the Artists Agency instigated the Writers Placement and Lucy often offered encouragement in my darkest hours. North Shields Fishermens Association provided not only the gear but moral support, and North Tyneside Council were brave enough to fund the book. The playwright Tom Hadaway gave me a deal of guidance with his vast knowledge of the fish quay. Al Hill's historical knowledge was always valuable, and he kindly wrote a foreword. The six boat skippers were remarkably open and frank with me, and I much appreciated their help. Thanks also to Jenny Attala of Northern Arts for the support, Eric Hollerton at North Shields Local Studies Centre for his help with my researches, Helen Sinden for design. Side Gallery Newcastle, and the People's Centre North Shields both helped with the photographic side, thanks.

Peter Mortimer's residency was assisted by Northern Arts through the Artists Agency, and this first edition of The Last of the Hunters, plus the residency have received financial assistance from the Government Urban Programme through North Tyneside Council's Inner Area Programme. The writing workshops during the residency received help from the Workers' Education Association.

BIOGRAPHICAL NOTES

PETER MORTIMER was born in Nottingham in 1943, and has lived in the North-East since 1970. He has now settled in the coastal village of Cullercoats. He has had four collections of poems published, and six of his stage plays have been produced in Britain, the USA and Germany. He is the founder and editor of IRON Press and magazine, one of the country's most active alternative publishers, and writes a weekly irreverent column for the Newcastle Evening Chronicle. In 1980 he was North Tyneside's first Writer in Residence.

By the Same Author

Poetry

Waiting for History
Utter Nonsense
The Shape of Bricks
The Ooosquidal
The Magic Castle

Plays

Snow White in the Black Lagoon
The Troutbeck Time Traveller
Imagine
The Man Who Played with Mice
IT
Elvis, Lucy & Capt. Sensible

FOREWORD

Fishing from the river Tyne and North Shields in particular, has been carried on for hundreds of years. Tynemouth Priory was built during the 7th century and it is known that the Monks from this Monastery were amongst the first to use the river as a base for their boats. By about the 11th century a number of "Sheles" (sheds) had been built for their fishermen, at a point near to where Cliffords Fort now stands, over the years the name for these huts changed to Shiels and then to Shields. Also at that time the Monks from Jarrow had built "Sheles" close to the river, this caused confusion to travellers and shipping to the Tyne, so that a distinction was made by preceding the word Shields with either North or South.

Round about the 18th century the fishermen moved with their families into "houses" built into the bank-side and overlooking the river. The word 'houses' is used loosely as they were nothing more than disease ridden hovels, connected by a labyrinth of stairs and passageways, with a public house at almost every corner. These public houses were used by the fishermen when they were ashore, they were also frequented by the Press gangs, and the fishermen suffered heavily at the hands of these bands of ruffians. It is not surprising therefore that a very close knit community developed over the years and a comradeship at sea that Peter Mortimer has portrayed so well.

Never in the long history of North Shields and Tynemouth has such an indepth study into the lives of the fishermen been undertaken and the findings of that study so comprehensively written.

Al Hill

The writer at sea — a damp note-book was often one of the practical problems.

Photo Keith Pattison

INTRODUCTION

In June 1986 I embarked upon what was to prove one of the most salutary experiences of my life — six months with the fishermen and fishing community of North Shields, the main fishing port between the Humber and Scotland.

My home, Cullercoats is only a few miles from Shields yet like most people my knowledge of fishermen was scant. Not only is it a tight-knit community, but when fishermen go to work, they disappear, like miners. Everyone knows mining is a dangerous trade. In fact the second highest mortality rate in the UK, with only one higher. Seamen and fishermen, where the chance of death is twice as great.

The bulk of this book concentrates on life at sea. Several writers have described life in the fishing community. The Shields playwright Tom Hadaway evokes it graphically in several plays. In 1986 there was a play on the Grimsby and Hull deep sea fishermen by Rupert Creed and Jim Hawkins, called "The Northern Trawl".

The difference between this book and those worthy creations is that this is written not from land, but from the sea itself. It soon became apparent during my six months' placement that while there were many aspects of shore life that were interesting down in low Shields, my main contribution, as a writer, would be in attempting to evoke the extraordinary life at sea on a small fishing boat, a life of such severity and hardship that few documentary makers, in whatever medium, have experienced it first hand.

It is a life which horrified me. It is a life which in turn affected my *whole* life, my sense of values, my appreciations, and which saw me return to shore from my trips, a different man. We know little enough of anyone else's trade or calling, but I soon realised this applied to fishing more than most. Within hours of being out in the North Sea, any folksy, romantic notions drain out through your unsteady legs, and you realise that this is a harsh battle for survival, that the North Shields fisherman often works grindingly long hours in conditions not tolerated by landlubbers for maybe a century.

They are at the mercy of the weather in many ways. At sea itself (and the North Sea is not known for its clemency) but also when prevented from sailing. No fishing, no money. It is a precarious life. The fisherman has little security and can be laid off at a trip's end without appeal. They are non-unionised (partly through their own resistance). They go to sea often with no firm idea when they will return. Many fishermen look older than they are, an affirmation of the work's demands.

In the six months I sailed on six boats, from the traditional inshore fishing of the North-East coble, to the long lonely journies of the Christine Nielsen, ten days searching for bounty. I wanted to experience all types of fishing, and these six more or less cover the field.

Thus on the Winaway, it was gill netting, on the Frem anchor bashing, on the Silver Coquet salmon fishing, on the Castle Dawn prawn trawling, on the Congener pair trawling, and on the Nielsen seine netting. All these terms are explained within. My boat trip chapters are in chronological order, to show any changing attitudes over the six months.

The book revolves around the main chapters of life on these different boats. Interspersed between are smaller sections springing from my talking to shore-based people, fishermen's wives, retired fishermen, net makers, herring curers et alia . . . I use the photographs of Nick Hedges who documented the Shields fishermen at sea some years ago (and whose work became an exhibition at the Side Gallery, Newcastle), the photographs of Tony Patterson, part of the People's Unemployed Centre's Photographic Project in Shields. Tony spent many long hours tracking down the people described in the book, and he also spent a day at sea. A third photographer, Keith Pattison came to sea with me on the Castledawn. He suffered for his art, but produced some memorable photographs. There's also the work of fisherman painter Micky Smith, Cullercoats artist Nicola Balfour, and the North-East artist Paul Rogers. And three "Fish Poems", putting, as it were, the other side.

Fishing — in North Shields at least — is an all-male preserve. Even in the quayside offices there's a predominance of men. It is a world where the devastation of existing fish stocks brings forecasts of future doom from many quarters. While not in any way wishing to detract from the seriousness of this threat (the herring and the mackerel were almost fished to extinction and cod could go the same way), my dilletante researches unearthed a petition against "a wondrychoun", a kind of beam-trawl which it was said "presses so hard on the bottom it destroys all life. In taking one good fish they destroy thousands for no real purpose". The petition was presented to Edward III in 1376.

It is a world which unlike farming, has few vested interests. Farmers have 60 MPs representing their interest, fishermen five. Fishing comes under the EEC Agriculture policy but receives only one per cent of the budget. In Britain average consumption of fish is only 5oz per week (meat 40oz), yet even small amounts of fish in the weekly diet can halve the risk of heart attacks (fish contains polyunsaturated fats).

There are few 'gentlemen fishermen'. It is a strongly working class pursuit, from the skippers downwards. In Shields it is still the preserve of the little man. Though boats are not always owned by the skippers, there are no signs of conglomerates or multi-nationals. Fish are still gutted, sorted, and stacked by hand. On the small boats the nets are shot by hand.

The Shields fish quay has changed little in the last 100 years (see the chapter on the New Quay That Never Was). In 1987 there's now a deal of housing and social development round the quay area, new chic dwellings, discos, renovated pubs and shops, night clubs. Just how much benefit this will bring the fisherman himself is doubtful. Where once there were hundreds of boats in Shields, now there are around 40. Some say the good days are gone forever, some claim sunshine is just round the corner again.

Who knows? Despite the advances of fish farming (now making inroads into white fish as well as salmon and trout), fish remains the only food which industrialised man goes out to hunt, which may partly explain our romantic associations. The hunt is a bizarre process where predator never meets prey until the chase is done, nor does he need to lay a hand on the victim to kill it.

The book's title comes from Lennie Myers, managing director of Caley Fisheries Ltd., one of the two Shields agents who runs the early morning auctions (themselves a unique experience) and keep the books and other financial arrangements of the boat owners (the other agent is Richard Irvine Ltd). I went to see Lennie on the first morning of my first day in the six months. He's a shrewd pragmatic man, and when I asked him why he spent his life involved with fishing, he paused for a moment and came up with the memorable phrase "they're the last of the hunters". I knew I had my title.

Where possible I've reported direct the comments and thoughts of the fishermen and others, though of course any criticism for the interpretation of my experiences rests with me alone. It is impossible to chronicle all the myriad activities that take place round the quay itself, an area of splendid, if often dilapidated Victorian buildings (some being renovated), a bustling little community splendidly free of such normal urban landmarks as banks, building societies, supermarkets, or buses. I hope the people selected will give some hint of its flavour, though if you're within easy reach may I urge you to go discover it for yourself. Better still, rise early and see the auction in action.

And while I resist strongly any cosy caricaturing of 'the fisherman', I must stress that I found them a unique breed. The ability to survive their environment, to return to it time and time over, to often make light of the most appalling adversities, made a deep impression on me. I found them hard men, yet rarely showing deliberate cruelty. My life to them was as strange and remote as theirs (initially) to me. The bridging of this culture gap was among the more interesting sideshows during the six months. Fishermen often hate being at sea yet are restless when on land. They can rarely plan a normal social life. Married life is stuttery. They often feel they must be mad to have chosen the way of life and many times during the six months I agreed with them. Yet back to the relative safety and comfort of the typewriter, and attempting to put the time into some perspective, I am now deeply grateful for having undergone such an experience. If the writing gives at least some indication of their working lives then I feel the book will have achieved something. It is not intended as a sociological document, but one writer's attempt to respond as truthfully and imaginatively as possible to a strange, often terrible, sometimes hypnotic world. The book is dedicated to the skippers and crew of the six boats on which I sailed.

Peter Mortimer
Cullercoats
April 1987

Shields' youngest skipper, Paul Dowse on the Winaway's slightly less than luxurious deck. *Photo Tony Patterson*

THE YOUNG MEN AND THE SEA

The Winaway

You approach your first trip on a fishing boat with trepidation. Norrie Morse of the Fisherman's Association presented me with a large black plastic bag; inside waterproof trousers and top, plastic gloves and sturdy wellington boots. Don these in front of your bedroom mirror and you can fool yourself you're already a fisherman. Norrie scrawled a good luck message on the bag, and at 4am one Summer morning I cycled from Cullercoats to Shields, down the long sweeping hill of Tanner's Bank where the pungent smell of fish suddenly invades the nostrils.

The Gut (the sheltered Shields harbour) is already busy at this time, the wee day boats chugging into life, small blue puff balls of diesel smoke as they turn into the Tyne and within half a mile leave behind the sheltering piers for the dubious delights of the North Sea.

It is a spectacular river to leave, Tynemouth Priory standing like a rotted tooth on the Northern side, also Tynemouth Castle, and the sturdy jetty built of Aberdeen granite. To the North the great beaches of King Edwards Bay and Tynemouth long sands, and further the silhouette of St. Marys Lighthouse, Whitley Bay; to the South, the empty sands of South Shields, down towards Roker and the distinctive red and white of Souter Point beacon.

Throughout these six months leaving and arriving back in The Tyne will be an emotional experience, and hardy though fishermen are, there are few who'd claim these moments leave them unmoved. A great love of the mainland sweeps over you, partly because of its distinctive coastline, partly because of the perils the sea offers, perils you are either about to face, or just return from.

My first trip is in the Winaway, a 37 foot stern trawler with the youngest skipper in Shields, Paul Dowse, 22. He has a crew of two, Kevin Dickinson 21, and Davey Carrick 17; add all three ages together and they don't amount to the age of some skippers I meet. The boat's now been adapted to gill netting, fine monofilament nets so thin and grey that when they lie folded in the boat they look like grandma's hair. Gill nets are shot to hang in great long walls down to the sea bottom. The fish can't see them, swim in and are trapped by their gills, struggling uselessly until they're hauled in, and they die. They don't die gently, the fine net often wraps itself deep inside the gills, sometimes in the mouth. It's ripped off, at times leaving a deep red gash. All sorts get trapped in the net; star fish useless for everything (apart from killing cats, who eat them and die of the poison), crabs, whose legs become so hopelessly entangled that they're snapped off, and the remains thrown on deck like some grotesque cornish pasty, sliding on its back, stumps still twitching; bad tempered cat fish with their dracula fangs, or sometimes something plain silly, like a pair of trousers.

But that's to come. We sail in the cold morning half-light, steaming 20 miles East of the Tyne. If you can, snatch some sleep during the journey, for there will be little else in the way of relaxation until 7pm. It's a small bobbing cork this boat, a denial of the computerised age, where almost every activity on board is carried out by hand, work little changed over generations. Like it always was, and probably always will be, it's hard, and it's dangerous.

Paul Dowse goes wrecking — plotting old wrecks on the sea bottom with the colour echo sounder in the wheelhouse, then shooting the nets above them. There's something ironic about catching fish at the scene of previous seamen's disasters, and there's also the danger of nets getting caught, and possibly lost in the tangles of the wrecks. But it's worth the risk; wrecks grow plankton which attract large shoals of fish. They also serve as protective hiding places, or at least they do until the fishing boats arrive.

The minute you leave the protective banks of the Tyne, you feel the growing swell. This happens no matter what the time of year, or conditions, an early warning that you're heading into one of the world's harshest seas, and you're doing so with very little between you and those elements. You do so knowing you may well fish all day and return with less than one box, wages that could be quoffed off in half an hour down the local; knowing that one lost or ripped net could wipe out more than a day's work and income. These are among the few things you *do* know when you head out to sea.

Seven nets, and the procedure is that they're shot last thing in the evening, left overnight, and next day you steam out, haul them in, one by one, then shoot them again. You do this as many times as the skipper thinks you should, but always they're left out there for the next day, marked on the charts and identified by the 'dans' (polystyrene blocks through which a bamboo cane is shoved and secured, a diamond shape flag fluttering at the top). Haul them in, and hope they're full of fish.

The Winaway is a small, open-deck boat. The wheelhouse is midships, a cramped affair with a leaking room, and one tiny shelf on which are crammed tea, coffee, and milk. For'ard, you drop down a hatch into a triangular shaped 'cabin', a gloomy dungeon with two bunks and little else, a leaking porthole, and an old plopping gas fire. Another hatch drops you down into the engine room, only three foot high, five feet wide, noisy, smelly. There's a toilet which doesn't work. That's about it — oh and the twin bobbin hauler over which the nets are hauled.

Boats of this size, (surprisingly given their operating area) are exempt from safety regulations governing larger craft; their skippers are also exempt from the normal captain's qualifications. I'd have preferred to have learned this knowledge on shore. Also midships is the exhaust funnel — wrap your gloved hands round for a few seconds comforting warmth against the sea winds (even in high summer this is a cold hunting ground).

Much of the work is done in a heavy swell and driving rain. Fishermen's waterproofs are just what they say — to a degree, but the wide sleeves and trouser bottoms (necessary for speedy dressing) often sees the salt water running up your sleeves, or a sudden large wave dashing against and fingering its way right up your boots. Chafing salt water is not one of life's great pleasures, and during the six months I was to see several nasty examples of salt water boils.

This is the one form of fishing where hauling the nets precedes shooting them. Left to the mercies of the North Sea overnight, nets can be destroyed or simply vanish, but on this day we locate the bravely fluttering flags of the dans, tiny markers in the huge swelling expanse of water, (Paul's chartings are good). The open deck has shallow sides, no higher than a man's calf. There are no rails to secure you. The dan is picked up, the net end fed through the boat's only visible mechanisation, the twin-bobbin hauler, a moderately efficient animal rendered redundant as soon as a net offers much resistance.

Even when hauled in without problems, Davey still needs to drag the net through, refolding it back in the deep plastic container. Fish hang from the net like sailors to the rigging; the thin twine often cuts deep into the flesh and Kevin's task, in rapidly disentangling them has to be without compassion, for already more net is being hauled.

This first sight of commercial fishing is a shock to the layman. There's often much blood in freeing the fish which are then tossed into a pile on deck to gasp and gape their great yawning mouths to death. They flick and flap, mainly cod, but also mackerel flashing their beautiful rainbow colours to the sun. For a few moments, fish see their hunter, they are put in touch with the creatures who invade their domain. Their mouths pout, fruitlessly gulping down yet more of that alien substance which is killing them. Their gills open and close like ventilators. Their eyes in death are no different to their eyes in life. A twitching, glistening, silent funeral pile grows on deck. Mass death is taking place before my eyes. With no other creature would it pass so unobserved.

Above us a screeching multitude of gulls await their pickings. When the whole net is hauled and folded, the skipper moves on to the next one, and the crew begin gutting. This too is a shock to the innocent eye, the belly sliced open, the foul smelling guts and intestines yanked out in one swift cutting and pulling movement, and the deck soon awash with gore and blood. As the innards are tossed or shovelled overboard, so the gulls descend, bad-tempered scrabbling and fighting. Birds are not the only creatures to feed off the boat. Humans too.

A few hardy 'leisure' fishermen track down boats such as the Winaway, hang about till we move off the wreck then move in, taking advantage of Paul's charting.

Preparing for the open sea — The Winaway about to leave the shelter of Shields "Gut". *Photo Tony Patterson*

This will happen further inshore, more than at this distance, but the anglers' lines and weights often foul the gill nets. Damage can run into hundreds of pounds in extremes, and each day the deck is littered with the angler's detritus, hooks, floats, weights entangled in, and ripping our nets.

On a future trip, an exasperated Paul yells angrily at a nearby leisure boat. In slightly less polite terms, the message is for them to go away, but they just sit there, awaiting our departure and there's nothing anyone can do.

"For them it's a hobby, a day out" says Paul, "for us it's a livelihood. Is it any wonder we hate the buggers."

Nets are shot from the stern. There's no mechanisation here, but there's a deal of danger. The first metal weight is thrown over the stern, dragging the first yards of net with it. Paul opens the throttle, the boat speeds away and the net whiplashes out, twisting and snaking off from the stern. Occasionally they need to guide it, reaching down into this violently moving mass of twine and net. There is little to secure oneself with here, and if the net (as can often happen) wraps itself round one of those guiding hands, it can yank the person into the sea. Fishermen have lost fingers as that angry thin twine lassooes them. On shore no activity of this nature would be tolerated without protective casing or grills. Nothing like that out here.

"I've been on boats since my father took me when I was five" says Paul. "I was in the merchant navy for three years, mainly P & O, but now I've got my own boat and I'm learning all the time. You go where you think the fish will be, but of course sometimes they're not. People think fishermen earn a lot of money. Sometimes we do. But this boat needs £250 to £300 a week just for running expenses."

On our first day, we took four boxes of fish, which would fetch around £45 per box in the auction. The money is split 50/50 between boat and crew. It's far from lavish.

This is as close as you can get to being a hunter. Every fish is pulled from the nets by hand, it's gutted by hand, boxed by hand. The sea is not some distant entity, but right there. Reach down your hand and touch it. Or it touches you, washing over you, spraying you. Days are spent in wet and damp. One minor hazard is keeping my note book dry. Someone had suggested bringing a camera, a tape recorder. The sea would sabotage either within minutes. Catching fish like this, you become reduced to essentials, you are self-contained. Even the dans are made by the crew from blocks of polystyrene and bamboo.

And consider circumstances when the net will not come up. This happens on our third net of the first day.

It will happen once again on the same day, a day when we haul up yard after yard of empty net, a routine of shooting and hauling that at times seems futile.

But more so when the sea refuses to surrender what we have sent down into it. Forty fathoms down, the net has ensnared itself on the wreck. A fierce tug-of-war ensues, Paul, Davey, Kevin (and feeble assistance from myself) dragging the net by hand inch-by-inch, often straining for several minutes only to see it snap back and render the effort useless.

We have 500 yards of net to pull up, the length of three football pitches and it fights us all the way. At the end, Davie's hands, encased in plastic gloves, are red raw, and as he says "nothing ever heals out here."

Inch by inch it comes, pulled over the hauler whose horsepower is totally inadequate for the task. Human muscle is all we have; the strain tilts the boat to port, as the net comes up, squeaking and straining, the pull pops out several of the plastic buoyancy corks. Heavily breathed obscenities intersperse the grunts. The net comes up full of sludge, full of rusty old iron, full of filth, empty of fish. The tension on the line is enough to break a limb or sever a finger were it to snap back suddenly. You pull to your limit, hampered by the deck's constant shifting; you pull never knowing at which moment the net may suddenly come free or snap. No fisherman will leave a net until it is totally impossible to salvage. It is £300, plus whatever (if any) fish are in it. There is the sense of the sea taking revenge, turning itself on those who come to plunder it. That romantic vision of fishermen pulling in the nets singing sea shanties looks far removed as the thin line cuts into your skin, you're 20 miles out in the North Sea on a rising swell fighting to drag a net 150 foot from the sea bottom on a tiny boat, and each pull leaves your body more drained.

This day begins at 4.15am. Through 17 hours the crew undergo extreme physical effort and the vagaries of the elements. There is between them a strange link. To call it romantic would be overblown by far; few words are exchanged as they work. Each haul is a struggle until the net is reclaimed. Obscenities are all you will hear. Yet there's something between them, from the moment the Winaway steams from the Tyne, and we are merely tiny specks on another tiny speck in an endless seascape. It is a mutual dependency. They do not talk about it, and why should they? It is simply there.

Davie says; "You tell your mates you're a fisherman and they think it's a nice job. I never swore till I did this. I do it all the time now. I go home and I swear at wor lass. It passes the time. Yes, I'd like to do something on shore sometimes, but I've only known fishing so far, and I'm going to become a qualified seaman. How many lads come out of a YTS scheme with a qualification? I will."

A few words on The Winaway's culinary customs. You cook on a tiny area little bigger than a chess board, two small gas rings, no working surface, one small shelf on which an elastic strap holds in tea, sugar etc. The matches are always damp, for there is nowhere on board which is dry. Strike enough of them and one will catch — the law of averages. There are four mugs, two with broken handles, all deeply stained with tanning; one small saucepan, a kettle, frying pan, rusting tin-opener, a water tap/pump that needs priming each time you use it, and then is maddeningly slow, sending a small trickle every few seconds, leaking as much as it supplies to your receptacle.

The boat lurches all the time, to combat which (partly) cups are placed on top of a bar towel. Food is either fried or straight out of tins. Either presents problems. Hot fat swishes this way and that, frying an egg slowly turns into

Preparing to shoot on the Winaway. *Photo Tony Patterson*
The right hand seaman holds the dan (marker bouy).

Winaway skipper Paul Dowse (left) helps mend a torn net. *Photo Tony Patterson*
The Co. Durham coastline is to the right.

scrambling it, and often you are sent reeling, hot frying pan in hand, to the far side of the wheelhouse (which of course is no great distance). Pouring tea or coffee necessitates holding the cup with one hand to stabilise it. This brings the risk of a sudden lurch and scalding water over your skin.

Unbelievably, in such conditions, one day Davie cooks *chips* (with curry sauce). The tiny sink is soon blocked solid with bits of food. It's like cooking on a see-saw, but less regular. It's like cooking on a camp-site except everything is moving about. Saving time and effort is important, so breakfast is eaten off a single plate. The first eats his, bacon and eggs, wipes clean the plate with bread, passes on the plate as the second portion comes up sizzling. A meal takes somewhere around 85 seconds to eat, then back on deck to work. What would you do here with a lunch break? Go shopping? A pint in the pub? The world measures only 37ft by 5ft.

On our second day we come across big fish. In the wheelhouse, Paul plays a bizarre sort of video game with the coloured echo sounder. He knows where the wrecks are, but pin-pointing them exactly and staying above them is a special skill. On the sounder they come up like columns of skyscrapers, yet too fast and you're past them, and a whole circle needed to come round again and you may miss again. A few feet here can make all the difference between a big shoal and an empty sea. And this time, it's big fish, up to 35 pounders (cod), fat beauties who lift all our spirits. Kevin frees them from the net, throws them on deck with a loud thud. Their tails beat against the wood, they struggle and bend, moving across the deck in a journey to nowhere.

But it's not as good as we thought. When they're gutted, the huge bellies slit open, it's to reveal the insides half-eaten away. Out of these bellies come pink luminescent worms, the length of your forearm. In one haul these worms have destroyed more than £50 worth of fish. Davie and Kevin smash them to pulp on deck, and with each blow, they shout "Bastards! Bastards!"

A special revenge is wreaked upon one of these monstrous wrigglers. It's dropped into an inch of disinfectant, and the crew observe the death agonies with a grim satisfaction.

And look what else we haul in. We bring up half a chair, we bring up a pair of fisherman's trousers (which Davie ties to the mast to dry — he'll be wearing them the next day). We bring up a World War II German machine gun, and when the barnacles are scraped off, the lettering is bright and clear, and the brass bullets gleam like new. And on this particular day, we bring up six and a half boxes of fish, a goodly catch of around £300. For the two days that follow the fishing will be disastrous, only half a box of fish on both the Wednesday and the Thursday.

Fishermen get sick of empty nets, and early on the Thursday Paul calls it a day. But the sea is to have the last laugh. The strong tides have dragged down one of the nets, the dan has been pulled under and there's no chance of finding it until the slack tide the following day. Paul's plan was to take in all the nets, forget fishing the next day. Now he has a long journey just to find this one net, and bring it home, the chances being it will have half a dozen fish in it.

Occasionally there's a lobster, not pink, as we see them in the restaurant, but a pre-boiled black. The crew fund their food and drink on board from lobster sales, though not all of it makes it home. Kevin picks it up, the giant claws positioned like a strangler's hands, a strangely mechanical life form. He flips it over and extracts a handful of roe from the lobster's underbelly.

"Want some?" he asks and shoves it in my direction. He swallows the roe, the lobster's claws click on. I have no stomach for it. On this, my first boat, my insides have regularly threatened to turn themselves outside in and downing the greasy food is less enjoyable than it might be in a domestic kitchen.

Great gulps of the astonishingly pure air serve as an antidote to the queaziness. I also eventually discover that the best (for me) solution is what in theory is worst — eating. Sea sickness on an empty stomach is pitifully awful.

Like all skippers Paul spends most time in the wheelhouse. Most of his conversation is not with his crew, but on the radio with other Shields boats. The language in these conversations singes the ears. It's not uncommon for the coast guard to complain of these broadcast obscenities, and further up the coast, the Northumberland fishermen think of their Shields counterparts as loud-mouthed and crude. Feminists would blanch at the verbal treatment their sisters are afforded on channel ten, and god help some saintly soul on shore inadvertently twiddling the knob to this frequency. Occasionally they talk about fishing.

For many skippers, radio may be the only contact with the other fishermen. Gossip, slander, hyperbole, abuse, camaraderie, all garnished with liberal four letter helpings. Don't take your granny on a Shields fishing boat.

The small boat steams back to Shields with its miserable catch. There's no fish room on boats this size, and the hauls are boxed and stored in a small hold. We stare at the one box — today's catch. Spending so long in such an environment for so little — the effort suddenly seems ludicrous.

The week-end lies ahead. On the return journey there's talk, some real, most fanciful, of what they'll do with their women folk. On this boat they're still young, unattached. They'll drink a lot at the week-end. Fishermen are tough, and Paul tells me that a few weeks previously he was done £70 for being drunk and disorderly. Being a fisherman, you need an antidote to work.

You can land fish at Shields from midnight onwards. For a boat such as the Winaway, which docks in the evening, this means leaving the catch on ice overnight, so that even on his day off, Paul Dowse will be down at the quay in the early hours. An early rise to land a few quids worth of fish, then back out into the North Sea to recover his lost net, and a few more quids worth of fish. Some day off. No wonder they fancy a few jars. I would too.

Colin Bonham — cheating death at sea. *Photo Tony Patterson*

DEATH AT SEA

The sea surrenders its riches to the fishermen. But occasionally it demands a price. Occasionally it takes back. All fishermen live close to death. It is part of the life. Sometimes they cheat it, sometimes not. This is a remarkable story of death, and survival, at sea. Colin Bonham is a 31 year old fisherman. One day in 1984 he and fellow fisherman Tim Stone took their coble out into the North Sea. Only Colin came back.

"Tim and I were almost the same age, only 11 days between us. His coble 'Dawn Marie' was named after his little daughter Marie. That morning, July 1, we set off from Blyth. I'd been in the building trade, but had come back to fishing for the previous 18 months. It was high season, and the salmon fishing was good, though it was a poor morning when we set off at 5.30am, wind straight from the North, perhaps force 5 or 6, a choppy sea. But it wasn't anything you'd worry about, fishermen go out in much worse, and we could see maybe another dozen boats around us. We moved South down towards Tynemouth for the first shot around 6am, hauled it in around 7.30am. It wasn't too good, maybe half a dozen salmon, so we decided to head further South, shoot again. That first haul was to be the last Tim ever made. We shot again, around 8am, and it was freshening by this time, but we were still in sight of boats, and I remember seeing Tim's brother Paris in his boat Rolling Stone. We were only two miles off the Tyne piers but the waves were beginning to get white tops on them.

When you're salmon fishing you move the boat along the net to take a look. We did this half an hour after shooting but with the wind and tide the nets had swung round. We decided to straighten them up. I was on the freedeck looking along the nets, Tim was on the tiller. I jumped down, grabbed the bobbin hauler, then I heard him shout 'Look out!' from behind. I had my back to the wind, I turned and there was this big wave coming broadside on. Now that wouldn't have been so bad, except it was breaking too. It picked up the boat and just pushed it along on its side so that we were 90 degrees to the water and filling up fast. I remember seeing Tim pull the tiller hard over trying to get us into the wind. The strange thing was we had a new lightweight tiller, not the traditional old oak one most cobles round here have. I think it snapped, and I think if it had been the old one it wouldn't have done. Suddenly I was down in the bottom of the boat and then I think I was thrown out. The second wave came and knocked the boat over on top of me.

The whole boat went under water, upside down, and I was inside it. But I could see daylight, I remember that. You don't have time to think; I just had to push myself out from under the boat and get up to the surface. There were bits of deck board and other debris floating around; I grabbed hold of them but they were useless, too light to take any weight.

I could see Tim, he was about 50 yards away and swimming towards a life-belt which was probably ten yards away from him. I really thought he would make it, but in that kind of sea you just disappear, and that was the last I ever saw him.

I kicked off my wellies then — they fill up with water and drag you down — but I still had on my oilskins and heavy jumper. I suppose I was threshing about for two minutes before I managed to grab hold of this plastic five gallon diesel drum. I was lucky; we had several of these drums on board, but this was the one I'd just topped the engine up with so it had only about two gallons in it and was buoyant. Did I have my gloves on . . .? Not sure, I think I did to start with, but lost them somewhere.

The bad thing was, while I was holding onto this drum, the diesel was slowly leaking out getting mixed with the water and getting into my mouth. It's the foullest taste you can imagine and I remember vomiting several times while I was hanging on, which was one of the worst things. I tried to shove the drum inside my oilskins but it wouldn't fit, so then I held the handle with one hand. I'd try to lay on top of it, or put my arms round it. Not very successful. All the time I could see the two piers of the Tyne and after I'd been in the water maybe half an hour I saw two boats. I shouted but no-one saw or heard me. The North Sea temperature is really cold even in Summer (it only changes two degrees throughout the year) and by this time the coble had gone down. There were bits of debris, I was just clinging on and I was freezing. I tried kicking out, but there was an ebb tide against me, and I couldn't make any progress. I could feel myself going numb, but knew I had to just hang on. I felt sure someone would spot me. The problem was Tim never used his radio to call into shore so nobody would have missed us, and there was no regular watch from the mouth of the Tyne. I could even see the coastguard station but knew no-one would be looking for us.

So I just had to hope. Hours seemed to pass. I saw two coasters coming into the Tyne but they were too far away. Being a Monday morning I'd had nothing to eat or drink that day. I was hungry, thirsty and when the waves broke on top of me I couldn't breathe. The only time I got warm was when I had a pee — funny that, enjoying peeing yourself. I never gave up, even though I seemed to be there all day. I lost track of time.

It was about 2.30pm, six hours after the boat capsized that Nat Herd came past in his boat The Frem. I saw him a quarter of a mile away and knew he was heading towards me. He fired off a flare, tooted his horn, and I knew I would be safe. I remember letting go of the drum seeing the crew on deck and taking the rope they threw me. Of course my limbs weren't in much of a state by this time.

No-one knew anything about us. We hadn't even been missed. Nat radioed ahead and by the time we docked in Shields there was an ambulance to take me to Tynemouth Infirmary where they stripped me off and wrapped me up in tin foil to get my body heat back. My blood pressure was so low said the doctors that I wouldn't have lasted much longer. I was shaking, all my limbs were stiff, I couldn't walk or anything. I was trussed up like a turkey for two hours, then they gave me a hot bath, something to eat, and they kept me in overnight. Meantime they were out looking for Tim but he was nowhere to be found. His body was hauled up some weeks later in the nets of the fisherman Tommy Taylor, at Souter Point, off Sunderland.

All this didn't stop me going back to sea. In fact I went out the day I left hospital, looking for Tim. And I started fishing again straight away. My wife Jean didn't try to stop me. And if it was the same sort of day again, same conditions, I'd go out again. It was just one of those things, it was a one-off, it could happen to any fisherman. You can't predict things like that, and I don't really think we could have done much to stop it. I'm back on prawn fishing now, on the Beverley Anne.

I can't really say whether or not there should be regular look-outs from the Tyne on days like that. The weather deteriorated I know, but there were some cobles like ours that were out all day, the same day. We were unlucky, that's all."

A brief escape from The Frem — skipper Nat Herd (right) enjoys a welcome pint in the New Dolphin on the fish quay. *Photo Tony Patterson*

FORGET THE TOILET BAG

The Frem

Nothing could have prepared me for The Frem. Over the six months different boats threw up different insights into fishing, my reaction mainly a mixture of astonishment at the way of life and admiration for the men who lived it. The nearest I came to hell was my time on The Frem.

She's a 64ft anchor seine netter, registered in Grimsby, built in Denmark, now at home in North Shields. Frem in Danish means "forward", and she herself took a runner from Frederickshaven Harbour the day war broke out. Two days later and she would never have made it.

At 50, she's pretty old for a fishing vessel. Her skipper's the same age. Nat Herd is a distinctive sight, long straggly dark hair, thinning on top and greying, a wild beard, front teeth missing; he's like some exiled '60's wild man of pop. He smokes 100 cigarettes a day. The Frem was once owned by his father, passed out of the family and has now partly returned to the fold, though the owner is Frank Molloy of Tyne Shipways Ltd., so whatever the Frem earns on her long lonely forays into the North Sea, Frank keeps half.

Study the boat, and the kind word that comes to mind is functional. Her open deck is like some giant meccano set, a mass of winches and winding gear, a stumbling obstacle course for any crew member. Giant bobbins hold more than three miles of rope which will shoot out, unprotected across the boat's rear half; more winding gear will pull up the giant chains and anchor from the prow. Catch any clothing or stray hair (page Nat) in this machinery and you'd be a gonner. The deck itself is slippery, stumbles and falls are regular. There's a sense of rust and peeling paint. The Frem is far removed from The Boat Show.

Normally the boat sails from Shields for journeys of ten or eleven days. I picked her up when she'd returned to port after a broken power block, half way through the trip. Only 24 hours into the journey I was to find some small comfort in the fact of my shortened spell.

We'd be catching mainly flatfish, plaice, lemon sole, witch, but also some cod and haddock. It's called anchor bashing; you drop an anchor, shoot out one and a half miles of rope, attach the net, continue, shoot out another one and a half, then journey in a giant triangle, returning to your anchor. The rope pulls along the muddy sea bottom (the slacker the tide the better) and the mucky clouds of mud drive the fish inwards. When you haul in the triangle shrinks, and the fish are increasingly driven towards the mouth of the 25 yard net. The rope costs £5,000, the nets are £1,000 apiece. Taking care of your gear is not just an act of love. Flatfish will sell (at this time) for between £18-60 a box, cod £55-65.

Nat will be after five hauls a day. This sounds simple enough. Until you see it in action.

There's a crew of two, 32 year old Archie Brunton with huge tattoed arms as ornate as a willow pattern plate. Archie's been fishing for 17 years, first left shore at seven, and is from a Cullercoats fishing family. One day he hopes to have his own boat (like several fishermen he has his skipper's ticket). Meantime he works on the Frem, looks around him as we sail from the Tyne and says "who'd believe how fishermen live."

Darren Shotton is 24, a Shields lad, that sort of massive but gentle power that Lenny had in John Steinbeck's Of Mice and Men. At one stage when I'm struggling two-handed with the bilge pump (and I'm not quite the eight stone weakling), Darren walks up, chats on, and takes over, quite unselfconsciously operating it at twice the speed one-handed. Fishermen are as tough as old boots.

Nat Herd is a quiet, shrewd man, who chooses his words carefully. We leave the Tyne at seven pm, a calm summer's evening, nine tons of ice on board. There's no refrigeration but the boat's cold dark fish hold will keep things frozen for the full trip.

"Look at it this way" he says "there's more fish coming out of a barren sea, than there ever was came out of a plentiful one. How? Because in the old days you had none of the modern navigational guides, you'd drop a grease covered line to the bottom just to find out if there was sand there, you'd spend half your time searching. Now electronic gear can take you within feet of where you want to go immediately."

In the wheelhouse, which again, doesn't really belong on a Boat Show stand there's radar, radio and a moving graph chart. Like most skippers he's plotted these waters over the years, and though fishing is always a gamble, knowledge helps. Recently the wheelhouse was broken into when moored up in the Gut. All his gear was stolen, and now there's an ugly, but necessary metal door.

A few homages to modernity then. Overall The Frem scarce stands comparison with conditions in British industry at the start of the Industrial Revolution. Climb down the ladder and you're in a 10ft 'cabin', no natural light, which will be sleeping and eating quarters for four people. There is no sink on The Frem. There is no toilet. There is no running hot water. The only tap is a cold one and it's positioned one foot above the floor. Washing is all but impossible and for the trip's duration (I stress that my five days was only half what they would normally endure) we all simply remain unwashed.

Clothes gradually stink more and more, stubble grows on chins, hair becomes lank and greasy. The filthy table cloth is littered with bits of food, dirty stained mugs.

Sleeping involves crawling into a coffin-cum-cupboard, a cramped dark claustrophobic place little wider than the human body. Reach up your hand a few inches from the lying position and you touch the wood above. This 'bunk' is about 5ft 6ins long which makes a full stretch and therefore, also untroubled sleep impossible, . Entry is made via a small sliding wooden door which then slides back to give the impression of premature burial. You are below the water line so that the sea rushes past you in great sweeping sounds only inches away. There is no light, and in my bunk the diesel stove close by produces nauseous, half-suffocating fumes. There is no pillow and one damp, blanket. There are no sheets. In addition to the sea sounds, beneath you, you can hear the water in the bilges sloshing constantly like some subterranean tide.

The horror of this confined space is difficult to communicate. Both Darren and Archie confess to occasional nightmares, waking suddenly to scream a long, frightened scream across the silent wastes of the North Sea. This 'accommodation' is at the boats prow, so that when we're underway you sense the water rushing past. This though I prefer to being at anchor overnight; the sense of utter loneliness and vulnerability to be bobbing on a dark sea, no-one on deck, all manner of fictitious beasties and enemies suggesting themselves with every creak and groan of the planking and machinery above, or more prosaically, yet no less frightening, the fear that some large tanker may miss our wee anchor lights and bear down on us.

"After a few days" says Archie "you get used to the smell, and anyway we're all smelling the same, so who notices. It's when you step ashore you realise how bad it is, when you see other people's reactions."

We steam from the Tyne for eight hours, weighing anchor at 4am, when we begin fishing. On this boat you work every minute of daylight. The working day is often a full 20 hours, hard physical work, unrelenting. It's a beautiful dawn and the huge anchor chains are run out across the deck, the two pink anchor pellets, looking like some comic clowns' prop for a circus, bob out on the ocean's swell. Nat shouts his instructions from the wheelhouse. Once the anchor's down, the spots are marked the rope is shot out and we're creating our first triangle. It takes 15 minutes to run out the ropes, then two and a half hours to haul them back in. On this first haul of the day these are the 'luxury' hours. On every other haul there will be fish to gut ice and box. This is the only spell of inaction. It's a matter of pride on the boat that no matter the size of the haul, it's all boxed before the next one is pulled up.

"I've seen us do 20 boxes between us" says Archie "not a fish left on deck when we hauled in the next lot." That's half a ton of fish, the two of them, gutted by hand on an exposed, heaving deck, in around two hours.

I'm never quite sure where we are on this trip. Nat keeps his position to himself. Nor do the crew know how many hauls they will take in what day, when they will finish. Nor do they know when they will sail home. All these things fishermen accept as part of the job.

"One thing about Nat" says Archie "he is a gentleman. Some skippers are real bastards, they make you feel like shit. I've never known Nat even shout at us. He works us hard but he's a gentleman."

The relationship between skippers and crew is often a complex one. Working so close brings an interdependency, a realisation of each man's worth. It can also create tensions, resentments. With Archie and Nat there's a small regular ritual to which both adhere faithfully, and which I'm sure in some way helps their working relationship. More of this later.

The previous year Nat sacked Darren for turning up late, then took him back on again. He could be sacked again. Fishermen have no protection against such measures, no union, no pensions, no redundancy pay or right of appeal.

The net is hauled in by hand over the hydraulic power block, a flotilla of seagulls bouncing in a white cloud behind us, anticipating the meal of the week. The bulging gut end is always a bizarre sight, a great swollen, wriggling pudding, gaping fish heads visible through the mesh, water running off in torrents. One tug of the knot, and the haul's free, splatting and splotting onto the open deck.

I'll discover here, as with other boats, that much of what is hauled in will be thrown back. Ginny skate, considered not worth the trouble are kicked out the scuppers. This is basic sorting, Archie and Darren plunging into the small fish mountain, kicking like chorus girls. Most of the fish thrown and kicked back are already dead. Look over the side and you spot their lifeless bodies, flashing silver underbellies sinking to the sea bottom, a brief and grotesquely vivid light show, growing weaker as the fish sink. A skate may perhaps survive a short time on deck, round fish, cod, and haddock are doomed, even were they thrown back immediately. Pulled up from as deep as 50 fathoms their guts swell horrendously, and when they're slit open the lungs resemble inflated condoms. Haul after haul, hundreds of ginny skate are killed.

The Frem's stern is so low in the water that even in calm seas (and this trip we were blessedly saved from any big swell) the deck is awash, each wave fingering its way round the deck then out through the scuppers.

Darren is also chef, working in a tiny cramped area, a gas stove wedged into a tin-plate lined recess. The gas stove has no regulator, and to prevent over-heating the door is wedged open with a tin can. Until this journey the hatch directly above leaked terribly and sea water would come splashing in on the hot fat. There is virtually no working surface, and cooking without sink or tap nearby is as frustrating as typing in boxing gloves. Despite this, somehow on the journey Darren will pluck from the fish ice a joint of lamb, stewing steak, a chicken, then (bizarrely), best steak for breakfast.

Cooking and washing up, plus disposal of the dirty water is such an obstacle, (clambering up a vertical ladder through a narrow hatch onto the deck), that all economies are made. Pots and pans are washed up in the water vegetables were boiled in.

Shooting the nets, hauling in, gutting, boxing. They stop for breakfast after which there is no real break. Being at sea on a small boat for such a length of time brings a permanent sense of tiredness, a semi-drugged feel.

"Everyone gets it" says Archie "When you get a trainee fisherman it's difficult keeping him awake." Add to this the hard toil of a day up to 20 hours, the terrible sleeping conditions, and it's small wonder that every spare moment on board (of which there aren't many) is spent either eating or sleeping. Eat, sleep and work, often with barely a word exchanged. The last haul is gutted at 10am, we eat in silence the day's main meal, then drag ourselves into our confined coffins to be called once more by Nat four hours later, wriggle out of the inhospitable bunk onto the cold dark deck, and without even a cup of tea the process begins again.

Darren says of it: "you really have to be psycho to live this life. Most of the time I'm at sea I'm really depressed, thinking how it's a waste of my life. It takes me days to get used to sleeping on this boat every time, eventually you just drop to sleep through exhaustion. Then when I'm on shore, and it's getting close to the time to go back to sea, I'm edgy, pacing about, looking for nothing in particular. Some day I'd like my own boat." At this his head drops and he grabs a few moments' sleep.

"Last coil!" It's the phrase no dozing fisherman likes to hear, the shout from deck each time the hauled-in net is getting close (the rope is split into seven coils). The three men haul along the counter-stern, a precarious lip of the boat with no supporting rails. Recently Archie clung on by the skin of his teeth when a large wave almost swept him overboard. As he says "I'd have been a gonner. No-one could have got me out of that sea."

The day seems insufferably long for a land lubber more accustomed to the civilised hours of shore-life. To me it is an horrendously freaky day, unnatural, yet a look in the official log (supplied for EEC statistics) reveals it as little different to most days. I pick a succession of six days at random. Hours worked are 20, 20, 19, 20, 21, 19, day after unrelenting day. Think of these hours, think of the working conditions, think of the environment, think of the danger. Then take a few moments finding another job in a modern industrial society with such a combination. Take your time.

When a cod-end full of fish lands on the deck the sound is like 100 fat ladies being massaged, big wobbly thighs slapped into shape. The plaice are as bug-eyed as Martians, the skate flip over to reveal strangely luscious kissible lips (on another boat a fisherman shows me how, in a somewhat distasteful piece of sexual gratification, those lips can be made to smile).

Grey masses of whelk spawn litter the deck like human brains. There are spider crabs, horribly beautiful, dog fish with their rough skin like grandad's stubble, the voracious cat fish which even when out the water will fasten its Hammer Horror teeth into your leg; there are monk fish with their great cavernous mouths twice the size of themselves. Each haul is a mass funeral.

"Social life?" Darren grins. "You just have to forget it. We never know when we're going so how can you make arrangements or anything. We just come out here and fish. Just have to have a good time if you can, on shore."

The doors on this boat are low enough to ensure only Ronnie Corbett wouldn't bump his head. Toilet arrangements are as follows; there's a deep blue bucket attached to a length of rope. Climb down into the roar of the engine room with the loo roll in your hand, squat on the bucket, and hope that the swell of the sea doesn't send you skidding round the floor bumping into the machinery, which it often does. The bucket is tall, and tapering to a narrow base, like a stove hat, and totally impracticable for the required purpose. When you've finished, climb back up the ladder onto deck, (taking care to keep the bucket well balanced . . .) throw it overboard, keeping hold of the rope, and wash it out. Take care the current doesn't pull it away from you. No-one bothers with all this rigmarole for peeing. Just point it over the side.

All ablutions are so awkward as to be put off to the last possible minute. Perhaps worst is the waking up in the night with a swollen bladder, a need firstly to squirm out of your bunk, clamber up onto a cold dark deck, slipping and sliding over the bits of fish gut. The pee sounds strangely loud, sounding out across those great black wastes.

Modern boats have shelter decks for protection, they have whale backs on the prow for more protection. The Frem has neither. There is virtually no protection from the elements during the working day. Only the small wheelhouse, wherein Nat smokes fag after fag, looking down on his crew and shouting instructions.

At night you lie in your bunk, cramped, dirty, tired, yearning for the small shorebased pleasures you'd previously taken for granted. The boat is a strange mixture. Below deck it's cramped, in the bunk especially the feeling can almost be one of panic, the claustrophobia such confines bring. To escape from this, you climb up on deck and stretching away from you on all sides is the vast emptiness, no sign of human life. Fish and birds apart, you see nothing, hour after hour, day after day. here aggrophobia can take hold, that terrible fear of open spaces, of a great, undefined nothingness.

Archie senses these feelings in me.

"Gets to you doesn't it?" he says. "I know it always gets to me. Leaving the Tyne always gets to me."

"I look at those two piers slowly sinking below the horizon and know that when I see them again, and the coastline of Tynemouth, Whitley Bay, South Shields, it will be a unique feeling. There's nothing like that sight for a returning fisherman."

There isn't either. On this trip, and others, my heart would fill with a huge longing and affection for my home base and activities, an avowal never again to take it for granted, a sense that I loved it passionately. A desire to touch your small possessions, to hug all your friends, merely to *be* there. To walk in a corner shop, to sit drinking in the local, or relax in an armchair — what great and unnattainable luxuries these suddenly seem!

Is this part of the secret, the reason why some fishermen endure what seems unendurable? Is it the contrast, that sense of enormous relief when you get on shore, and the chance perhaps, even though briefly to live life back there more intensely than those who never leave it, for it is difficult to communicate your feelings to those normal land-locked souls? Is it any wonder that fishermen often drink when they return (I never saw one touch a drop at sea)?

"I go out into Shields when I get back" says Archie. "I spend maybe £60 or

£70 in one night, so what? I've been on this stinking boat for ten days. But look, I'm not moaning on. As a fisherman you take what you get. Some boats in Shields are better than this. Others are more or less the same. That's how it is."

Darren is less charitable. "I reckon it's a disgrace the way we live. Who else would put up with it?"

"Some people go crazy. And I often feel like it. I knew this bloke, he'd had such a bad time on board a boat that he pulled on a life jacket when he saw another boat passing fairly close, jumped overboard and tried to swim to it. No-one ever saw him again."

We plie the North Sea for five days, the haul totals slowly rising. We'd all like to go home with 50 boxes.

Archie joined The Frem three months previously. Then he weighed 13 stone. In that time the work has seen him lose one and a half stone. Like most fishermen he is deep chested, broad, a concentrated strength and power. "Never pick a fight with a fisherman" he grins. I don't need the advice, but I gladly pass it on.

By the end of the fourth day, having done some labouring, but nothing like the work load of the crew, I am utterly exhausted, filthy, depressed, and bewildered, a kind of culture shock. Being a writer, which I'd previously considered a precarious following on its own terms, suddenly seems paradise. To sit, clean, refreshed and warm in front of a typewriter, with a floor that *doesn't move around*! My clean wee toilet bag has sat neglected. Sleeping and walking in the same clothes as you've worked in no longer matters. Those twice-daily scrubs on shore here seem like cleanliness taken to obsession. At night in my bunk the only interruption to the darkness is the series of strange shadowy shapes thrown up by the flickering diesel stove.

During the restless sleeping and constant half-waking, these appear as demons, sea monsters, dark creatures of the imagination. Darren's words about men going mad return to my mind.

Four full days we fish, in that time working more hours than most people in a fortnight. Night is falling on the fourth day, the rain is sleeting down, the last fish is gutted and all thoughts turn towards the long steam home, and dry land.

"We'll do another shot" says Nat. It's unbelievable. This means another three hours work in dreadful conditions, wind, rain and tiredness to combat. Yet a good haul can bring in £250 worth of fish, and good skippers are loathe to throw up the opportunity once they're out there. The crew say nothing, not a word of complaint, at a time I would expect a mutiny. Unable to take any more on this open deck, I descend to the dubious delights of the cabin, close my eyes against the realities.

And so to the small ritual I mentioned. As the trip progresses Archie keeps the box tally, and each evening as we eat the late meal informs Nat of the daily, and running totals. There are two totals, the official 'book' total which Archie announces, and the higher, actual one. This is nothing to do with deceit, both men are aware of the ritual, both men respect it. When we are finally steaming back to the Tyne, Archie will announce the true total, several boxes up on the 'expected' one, thus giving a small fillip to the skipper. We have reached our target of 50 boxes, which will fetch around £2,000 in the early morning Shields auction. From this Archie and Darren will take home around £200, and Nat something more than that (he wouldn't give me an exact figure). At a quick look it's not bad money. A closer one tells a different story.

"And remember this" says Nat. "Every fisherman at sea keeps eight men on shore in work."

We will land in the early hours. Eyes scan the horizon for the first sight of Tyneside.

"Back to Shields" says Nat, about 500 cigarettes the better (or worse) since we set sail. "The port's in decline" (though '86, with rising prices and falling fuel costs has been a good year). "Ninety per cent of Scottish fishermen now boycott it because some years ago Shields buyers got too greedy. They kept the prices artificially low when the big Scottish catches came in. Now we've been left behind, and Scotland's the place. What a pity. Shields is a great natural port. Fifteen years ago they planned a new quay here. Look what happened. Nothing."

Darren nicked off school early to go to sea. He lied about his age to the first skipper, and spewed up on his first trip.

Times on shore are precious but sometimes short. Darren says: "We once got into Shields at 11.30pm, the skipper turned round and announced that we were sailing again at midnight. I rushed home in a taxi, had a shower and came back."

The Frem looks a laughably exposed and vulnerable craft to be undertaking such journeys. She's been doing them for 50 years, she's been out on those seas since Edward VIII left the throne, since the start of the Spanish Civil War. She's been out there as long as the world's had television broadcasts, and since the old Queen Mary's maiden voyage (a ship she's outlasted).

On the return journey, the decks are washed, machinery put away, the nets are folded. And for the final few hours we crawl back into our coffins, the luxury of sleep without waking for work.

I wonder about this boat in bad weather. Even in Summer the low sides see waves regularly looping over the top, and despite the waterproofs which are worn every waking hour, the crew get wet and cold. It's like the most grinding work in a factory, except the factory floor keeps throwing you around and there's no roof, no walls, no heating. My legs are covered with bruises, a slippery deck often tossing you against the metal machinery. Where else would men work 20 hours a day without toilet or washroom, without hot water, without, more or less, *anything*?

They'll return, briefly, to wives, girl-friends, families. I step from the boat in the early hours, 'phone a taxi from the harbour-master's cabin. I climb in, and see the driver wrinkles his nose. My clothes could be peeled off. Dirt and fish layer my skin. I swear, at this moment, that one million pounds would not get me back on The Frem. The crew will return for much less.

The Fish Quay by Nicola Balfour

"Being married and not married"
Mary Morse, erstwhile fisherman's wife.
Photo Tony Patterson

FISHERMEN'S WIVES

Marriage to a fisherman can bring fear and loneliness. Mary Morse's marriage to Allan didn't survive. Helen Crosland's has.

MARY MORSE

"It's like being married but not being married, such a strange life, and I think that's why fishermen's wives stick together. They're very clannish, and don't make friends with many outsiders. They understand one another, and they also understand that when a man's at home, you don't contact his wife, because that time is precious to them. Sometimes it feels like being a squirrel; you have to hoard the money against the bad times. My husband was at sea, and also my sons. It's funny you worry more about the sons. You'd watch for the weather forecast. A normal marriage? I'd probably have found that hard to handle! You adapt to him being there then not being there. He comes in all tired and mucky and stinking of fish but that smell is worth ten guineas an ounce because it means he's home, and he's safe. Then after a bath, and a couple of pints he's OK, but if the boat doesn't sail again when planned, he becomes agitated, uneasy. It's the sea, it calls them back and however bad it is, he'll be raring to go when the time comes, all wound up. I've got three sons, one is a fisherman. Of the other two Norman tried it for seven months but it wasn't for him, Steve has never been out and has been warned off the life by my eldest son Allan, who is a fisherman himself."

Helen Crosland. *Photo Tony Patterson*

HELEN CROSLAND

"I've been married to Aidan nine years, and we've been together for 14 in all. Often you feel like a one-parent family, though I would never change it for anything. Aidan took our nine year old son Aidan out for a week and he loved it. I think he'll be a fisherman himself. Sometimes when Aidan Snr is home he gets under my feet because I'm more used to being in the house on my own now, or just with the two kids. When your husband's away like that, you have to make sure you're on good terms when he leaves, and we always do. Of course the night before he sails he can't sleep. All fishermen are the same, on edge.

Of course it affects the bairns, their father comes home, they get used to having him around, then he's off again. Then you work at the house, get it really spotless. In he walks after a week at sea and he's in a filthy state. But you get to be fairly self-sufficient, and you get to be lonely too. You try to cram everything into a few days when he's ashore. I'm all over him when he comes home."

Keith Bowie — end of an era for salmon fishermen? Photo Tony Patterson

ALL WORK NO PAY

The Silver Coquet

Imagine rising with the dawn, going off to work alone, in a dangerous environment, spending £10 in expenses, and returning home without a penny. Some job.

Not that this would be a typical day for a coble fisherman, but it does happen, and it happened to Keith Bowie while I was with him. Keith is a 35 year old inshore fisherman. A salmon licence is a coveted possession in these parts; only 75 to be had (they cost £400 apiece) and you might wait years to get one. This scarcity leads to all manner of tales on the money earned by the salmon fishermen.

"The truth is" says Keith "this kind of fishing is almost over. This could be our last year".

What's pole-axing the small coblemen is the rapid rise of salmon fish farming which has sent the price of the fish plunging by up to 75 per cent in one year. In 1986 10,000 tons of salmon were farmed, 3,000 tons up on the previous year and with such giants as Unilever and Marine Harvest Ltd now involved, the future for the small man looks dim.

If this were chicken and eggs, then there would be all manner of 'green' parties arguing for free range as against factory farms. For some reason the argument has never been taken up as regards salmon, and it's doubtful if anyone can tell the difference between salmon caught on the high seas, and those nurtured in a farm.

Also remember the new salmon bill severely restricts the North East drift net fishery and is a serious injustice in the view of inshore men."

The coble is the traditional North-East boat, flat bottomed for stability, shaped to be launched direct from the beach, tarpaulin stretched like a tent across the bow, and with a long distinctive tiller handle the size of an elephant's trunk. They're romantic looking craft and fast vanishing. At one time 60 cobles would fish out of Cullercoats; now there are four.

Keith Bowie steams a mile offshore, and shoots 600 yards of net which hang close to the surface where salmon swim. At this time of the year — August — they're returning to their spawning grounds, running North-South, and the nets are shot to cross their path.

On a really good day, Keith might make £400 — his best tally was 46 salmon, in 1986. On a bad day — and this is a bad day — he will catch nothing, several hours running along the nets without a sign of those red fish (the name many superstitious fishermen call it — believing the word 'salmon' to be bad luck).

Apart from the diesel engine, the only mechanisation, indeed the only change to a coble from its original design is the bobbin hauler, fastened for'ard on the port side. Even this is frowned on by some old timers who still shoot and haul the nets by hand, believing the hauler's weight upsets the boat's delicate balance.

The limit is six miles, and there's a one mile 'box' round the mouth of the Tyne, like all rivers. Cobles offer virtually no protection and although the design gives them a "grip" in the water, and they are built to be launched in the face of breaking seas, one large wave could capsize them. Keith was a lifelong friend of Tim Stone (see 'Death At Sea') whose own coble fell victim to one such wave.

"It's become a rat race you see" says Keith, as we run the length of the nets without a fish in sight. "That day Tim died — I don't think many boats would have been out, one time, but it's getting so as there's no such thing as a bad sea anymore. People are having to go out in almost any weather, due to economics. The seas were very heavy that day".

On this day it's a medium swell, enough to remind you of the vulnerability of such a small craft. And fishing when there are no fish to be caught can be a tedious affair. Every half hour we run the length of the nets. We nibble on sandwiches, chat. There are no fish. Keith opens up the throttle runs along the nets at top speed, tries to scare them in. There is none to scare.

"Sometimes I wonder why I do it. A lot of fishermen are getting out of salmon, going on the trawlers, and who can blame them?"

North Shields once built its own cobles — back in the '30's. "The Northumbrian coble, nicknamed the "arrow" is, to my eye, a much finer looking boat than the cumbersome Yorkshire variety (called 'tanks' up here). You see some cobles with a wheelhouse attached, but to me they lose their lines".

Astoundingly, Keith Bowie can't swim. Like many fishermen he's philosophical about it, believing the North Sea wouldn't give you many chances however good a swimmer you were.

This is the most basic fishing I encountered. There's no radar, no sonar gear. The fish finding is a matter of hunches and the nets are shot across a length of bamboo pole, spitting out mud as they twist away.

Economically, fish farming makes more sense. But the sight of the North-East coble is one we'd sorely miss. Keith's boat (which he skippers, owning a cornish cove boat named after his mother) is called "Silver Coquet" — a Northumbrian river.

"Cullercoats was once the largest fishing village on the North East coast" says Keith. "There is no better illustration of the decline in inshore fishing".

THE FISH POEMS

Most of my response to the fishing trips was via the documentary prose approach. But being a poet too, I also found the odd verse wriggling through. Three of these seemed moderately interesting. I call them The Fish Poems because two out of the three speak with the voice of a fish, a viewpoint not often put in this industry.

A FISH SPEAKS

Imagine we had voice
Imagine our cries of pain
as you lift us from the sea
drop us to rainbowed death
on your decks.
Imagine our multitudes
shrieking.

But our Al Jolson mouths
are silent.
They open and close
like songless choirboys.
Our blinkless eyes stare
before dying
at your swaying heavens.
Our tails flick like ladies' fans;
mother-of-pearl brooches —
our scales in the sun.

Pile us deep to die
no sound will disturb your ear.
We are no louder than
the sinking drowning man.

THE FISHERMAN'S HOME

Beneath his feet
the sea arches its spine,
writhes and twists;
a bucking bronco
he sometimes calls his own.

Quicker than the sea
the strong fisherman tires,
pulls in his nets.
Salt water boils
crack open his skin.
Still quick, he returns
to garden sheds, to perfume
to letters from the bank.

Where others walk steady
he feels the soil and concrete
shift and heave.
This neither is his home.
Cracked pavements tilt him
nudge him back
to those awful wastes.

THE FIRST FISH PLAN FOR THE FUTURE

Rule number one — we avoid man.
We will live in cold oceans
where man cannot breathe,
far from sun and air.
We will live without sound
so man cannot hear us.
We will live without light
so man cannot see us.
We shall be quick and nimble
so man's clumsy land limbs
cannot follow us.
We shall be thin and slippery;
man's ugly hands cannot grip us.
We shall stay hidden from man's world
Man will quickly forget us
We shall cause man no anger
Why should man bother us?

Dive deep, my children
and we shall live in peace.

John Carter (centre with notepad) auctions more fish. "I'm an aggressive seller" he says. Photo Tony Patterson

"The floor is carpeted with fish" — early morning at the quayside auction. Photo Nick Hedges

THE DAILY FISH MARKET

Long before most of us are awake, the daily auction market at Shields fish quay is underway. Boats land their catch anytime from midnight on, and the boxed fish are laid out by the 'lumpers'. By the time the auction begins (7.30am) the ground is carpeted with fish. The boxes stretch away (more than one thousand on a good day), serried ranks of blank eyes stare up. There's cod, whiting, plaice, witch, cat fish, dog fish, monk, lemon, haddock, mackerel, and they're laid out in a vast echoey shed longer than an ocean liner. One side is enclosed, the other opens up to the river and the sea.

Huddles of merchants move round the different catches, printed markers identifying the boat. The merchants have markers too, and once they've bought a consignment from one of the auctioneers (there are six in all, three from Caley's, three from Irvin's, though they don't all work at the same time), another marker is slapped on a convenient corpse in the 7 stone box. The boxes are then dragged off by men with long metal crooks. There are fork lift trucks too, and though to the layman the scene is one of noisy chaos, the well-oiled efficiency is such that within an hour all the fish is sold and disposed of.

Auctioneers have different techniques. Suddenly one will stand on top of a fish box and incant; it's like the mouthings of religious zealots, and once the incantation begins, it attracts a crowd of dealers. It's hard to see just who is bidding (dealers have their own individual signs, known to the auctioneers) just as the language might seem mystifying to an outsider.

"28 bid the witches", "60 the cod" — the auctioneers words seem to be spoken to the air, the knots of serious looking rubber-booted men show few signs of movement. Mainly they stare down at the boxes. Yet suddenly the sale is made, the auctioneer makes a mark in his book and moves on a few feet, pitches camp round some different boxes and the process is repeated. The huge carpet of fish boxes shrinks like a melting iceberg, men in wide rubber clown trousers move in, and within minutes they're loaded into the vans and wagons of the wholesalers or retailers (there's no differentiation made here).

Like most of the fishing industry, this is a man's world; of the 50 or 60 dealers only one woman is on view. Ice skids across the wet floor, at the edge of the shed, gulls wheel and screech, waiting for pickings.

One of the regular auctioneers, John Carter has a sharp urgent voice which almost insists you make a bid, daring you not to. Another, Bill Ponton seems to incant, leaning back on his heels, prices quoted with half-closed eyes, a guttural sound from somewhere down in the throat. George Harvey of Irvins sells fish like a playful uncle, a half-smile always at the edge of his lips.

Shields is a natural harbour, with 24 hour access. Traditionally fishermen have often mistrusted the dealers, suspecting them of operating 'rings'. Dealers aren't always over-enamoured with fishermen either, but every day, the auction takes place, the fish get bought and moved out.

George Harvey (white coat) selling fish "like a playful uncle".

THE NEW QUAY THAT NEVER WAS

Not much has changed at North Shields fish quay in the last 100 years. That may be a nostalgic thought for conservationists, but it's been more a case of muddled planning, half-hearted commitment and unfulfilled promises. While other main British ports have seen money poured into them, Shields has been relatively neglected, and the Great White Hope of 1972, the new quay has, like many previous grand schemes for this area, been stillborn. Many Shields fishermen are still bitter about this recent history, and recounting the tale may shed some light on how it came about, or in this case, didn't.

The fish quay was taken over in 1968 from the Tyne Improvement Commission by the Port of Tyne Authority — with reluctance by the PTA some said. The next year the Port of Tyne produced a report suggesting the quay be moved to the nearby Albert Edward Dock as improvements to the existing facilities would cost £½m. It is the biggest port between Humber and Aberdeen, in a delapidated condition, and most parties agreed, was "ripe for expansion".

This move never happened and by 1970 the present quay's future looked so gloomy as to prompt talk of closure itself. By 1971 people were realising how outdated were the facilities; in the words of one observer "holding back the growth and expansion of what is potentially one of the most important fishing ports in the country".

Everyone agreed, something must be done. In 1972, by which time visiting fishermen were being discouraged not only by the facilities, but by the rapidly increasing landing fees, a £5m scheme for an entire new dock was launched, a modernisation which according to the Shields Weekly News would "take North Shields from the '70's to the '80's". The scheme was announced simultaneous to news of government grants of 60 per cent for such dock extension, meaning £1m would have to be raised locally. The Port of Tyne at this stage would be expected to find an extra £160,000 a year towards running the new dock, planned beneath Knotts Flats, east of the present quay.

By March 1973, the PTA had still not formally applied to government for this grant. March was the deadline, they asked for an extension, and by September the Authority was announcing that "the go-ahead is expected". Months of talks followed between the then Tynemouth Corporation (who owned the land) and the PTA (who would lease it).

"All the parties concerned are very enthusiastic" said Councillor G. Bilclough of the Trades & Commerce committee, adding "we have got the impression the scheme will definitely go ahead, no doubt in my mind that the PTA will build the dock in the not too distant future". At this stage it seemed the government grant was the only remaining obstacle.

By October of the same year, prospects were splendid, so much so that firms such as Boston Deep Sea Fishing opened a Shields office in anticipation of the new dock. Designers were appointed in December, and in Spring '74 contractors were invited to tender for the work to build a 2,200ft breakwater, with 2,900 sq feet of concrete quay walls. Work would start that year, and finish late in 1976. The government, though still with no official announcement, was making the right noises, and plans were so well on that there was talk of turning the old quay into a marina.

Tynemouth's Tory MP Neville Trotter was becoming a regular questioner in the House of Commons, and in June 1974 his question to the Minister of Agriculture Norman Buchanan brought the first veiled warning that government money may not simply be a rubber stamp. By October the same minister was making even more ominous noises, mentioning spiralling costs and saying "we shall have to do our sums again".

Plans were for a quay catering for two trawlers and 24 fishing boats at any one time landing fish. Waiting and lying up berths would provide safe accommodation for four trawlers and 80 seine net boats.

And in 1975 it seemed it would indeed go ahead, with the announcement from Christopher Soames, then vice-chairman of the EEC that the community's agricultural fund could offer a 25 per cent grant, which coupled with government money would mean only 15 per cent had to be raised locally. This was in the Spring, though on Tyneside at this time, there were suspicions that the Port of Tyne's commitment to the new quay was not all it should be. Council leaders, and letter writers to the local press urged PTA to stir themselves and help push the scheme through.

With neither the government nor the PTA taking a strong lead, the scheme's precarious nature again seemed evident. There was a new fishing minister, John Silkin, and by now he was replying to Neville Trotter's questions in the best tradition of political double-speak by saying "this is a matter I shall consider in relation to the future of other ports".

His 'consideration' was such that the government killed the scheme dead within the next two months (rumours were that the initial cost of £5m had now doubled to £10m), and in a gesture which some saw as rubbing salt in the wound, the government offered £250,000 to make repairs to the existing quay; this at a time when other ports such as Lowestoft, Grimsby, Fleetwood and Hull, had been given grants of around £1m — some more than £2m.

The sense of outrage in Shields led to huge campaigns, petitions, but to no avail. In 1978 Tyne & Wear Council tried for a £10m Common Market loan, but this was rejected. In May of that year an all-party committee of MPs, including Neville Trotter issued a report. It called the present quay "dangerous, primitive and incapable of expansion", and launched a scathing attack on the Port of Tyne Authority.

The report, which has never been acted upon, suggested a new quay should be built, to be run jointly by North Tyneside and Tyne & Wear councils (the latter of course now no longer exists), and it said the PTA "lack the necessary enthusiasm and will" to undertake the right plans. Tyne & Wear attempted to wrest control from the PTA the same year, without success, and the fishermen found themselves now facing an almost impossible task in reviving plans for the new quay. Local Councillor Jim Bamburgh summed up many people's feelings when he said "the government is condemning Tyneside industry to death".

Dissenting arguments were rare, but one came from local fish merchant Dennis Bell who called the planned quay a waste of money. "What's the use of big new premises if the fish aren't available to fill them?" he asked.

With almost every industry in decline on the Tyne, an infusion of life into fishing seemed to many the best way to revive the economic fortunes of the river. That chance it seems has gone forever.

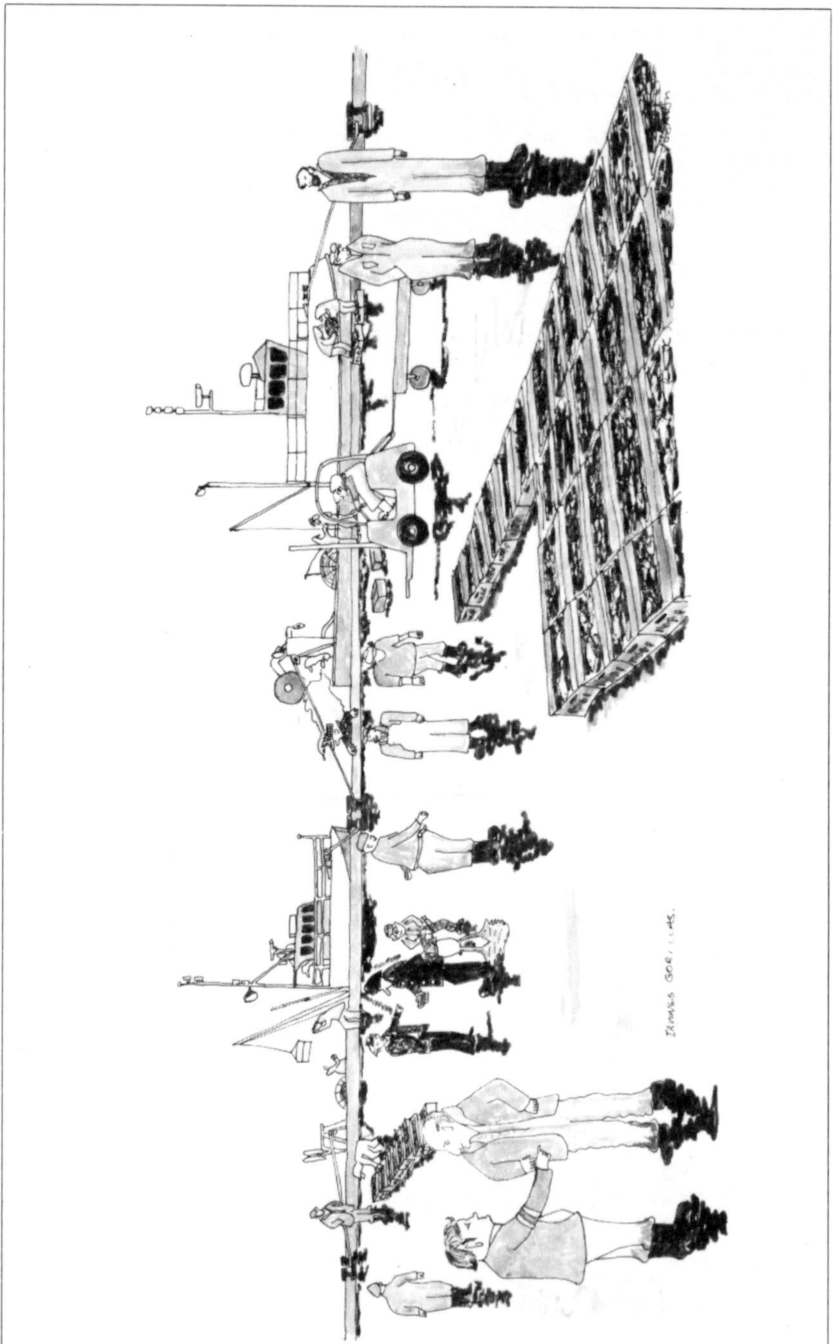

MICKY SMITH

Micky Smith was born in 1946 and has been a fulltime fisherman since 1977, at present working aboard the 60ft prawn trawler "Luc", out of North Shields. He's painted and drawn since childhood, is mainly self-taught, and usually depicts the fishing industry in his work, either Shields boats (Christine Nielsen, Condowan, Tudor Times etc), or caricatures of fish quay life. He hopes to put together a collection of his drawings for a future book. Here he writes his own descriptions to the cartoons.

IRVIN'S GORILLAS

"The 'Con' boats, the Congener and the Camelia are both agented to Irvins and are seen landing their catches onto the quay for Irvin's lumpers to lay out. Alan and Tony appear to be discussing what to lay out first while other members of the team are watching for Davy to bring the boxes over with the fork lift.

The boats' owners, 'Old Alan and Young Alan (Morse) are seen reminiscing about the ones that got away, or perhaps anticipating a poor market. The quay master is drawing the local constable's attention to that damned night reveller who has sneaked back onto the quay, and a kindly crew member of the Congener is helping a Marine Division Sergeant to remove a seagull's droppings from his shoulder."

ANOTHER HEAVY NIGHT FOR CALEYS

ANOTHER HEAVY NIGHT FOR CALEYS

"The cartoon refers to a mid-Summer Friday morning down on the fish quay. It was a very heavy night for landing fish with something in the region of three to four thousand boxes coming in on the boats, all to be laid out before the bell went for the start of selling at 7.30am. Caley's were responsible for about 70 per cent of this.

All night boats were coming in, one after another as Caley's lumpers were landing between 200 and 400 boxes from each boat. It was a particularly warm morning, the sun was high in the sky and Caley's men were sweating and exhausted. They were relaxing (or collapsing) on the trolleys while Nick (Mr. Tetley) was drinking his first cup of tea of the night. Cyril was allegedly taking oxygen when at about 6.45am much to their consternation one of them noticed the 'Novospere' and 'Stardust' charging up the river towards the fish quay. These were Caley contracted pair trawlers with another 700 boxes of fish between them to land and lay out before the 7.30 bell.

Enough to say that the skippers of both boats were definitely not the most popular of people among Caley's lumpers that morning. But between both sets of lumpers (Caley's and Irvin's) the boats were landed before and during the market, the last tiers of boxes going down just as the auctioneer was selling them."

THE COUNT

"Each fishing day the crews of the prawn trawlers after having sorted their catches land them by derrick onto bogies on the quay. These bogies are then raced by the crews along the quay to the agent's representatives to be weighed. As most of the boats land at roughly the same time a serious bottleneck often develops on the scales. Because of this skippers can be seen and heard giving 'friendly advice and encouragement' to their crews. The crews cheerfully trundle the bogies towards the scales trying to sneak in front of the boats who have landed before them. Someone (probably the apprentice) is sent ahead to raid the buyers' wagon, to steal the empty plastic boxes to put the next day's catch in. A good apprentice should be able to keep his boat stocked with at least two days' supply of boxes. It's no good having prawns on board and nothing to put them into.

When the bogies are formed (by fair means or foul) into a queue, the boxes of prawns are passed to the agent's representative who assists the skipper of each boat to weigh them off into their three stone units. Prawns are added to or taken off each box (amid much argument) until an exact measure is ensured. The accuracy of the scales is often discussed, over the radio channels or the 'steam-off' each morning.

The three stone units are then passed to the buyers' representative who accurately (some skippers think too accurately) weighs a small amount of prawns and takes a sample count. Upon the count (that is the number of prawns to the pound) depends the rate paid per box.

This leads to a variety of expressions around the counting area. 'Twenty Two's' or more mostly brings a groan or a straight 'Take it again', whereas 'Sixteen's' can often bring out a sly sneaky smile and a rub of hands.

The buyer's representative is very keen on the count and concentrates very hard, not allowing anything to disturb him, although he has been seen open mouthed and fumbling at the count of 'Three's' — *'Bugger me, theor bloody lobsters'.*"

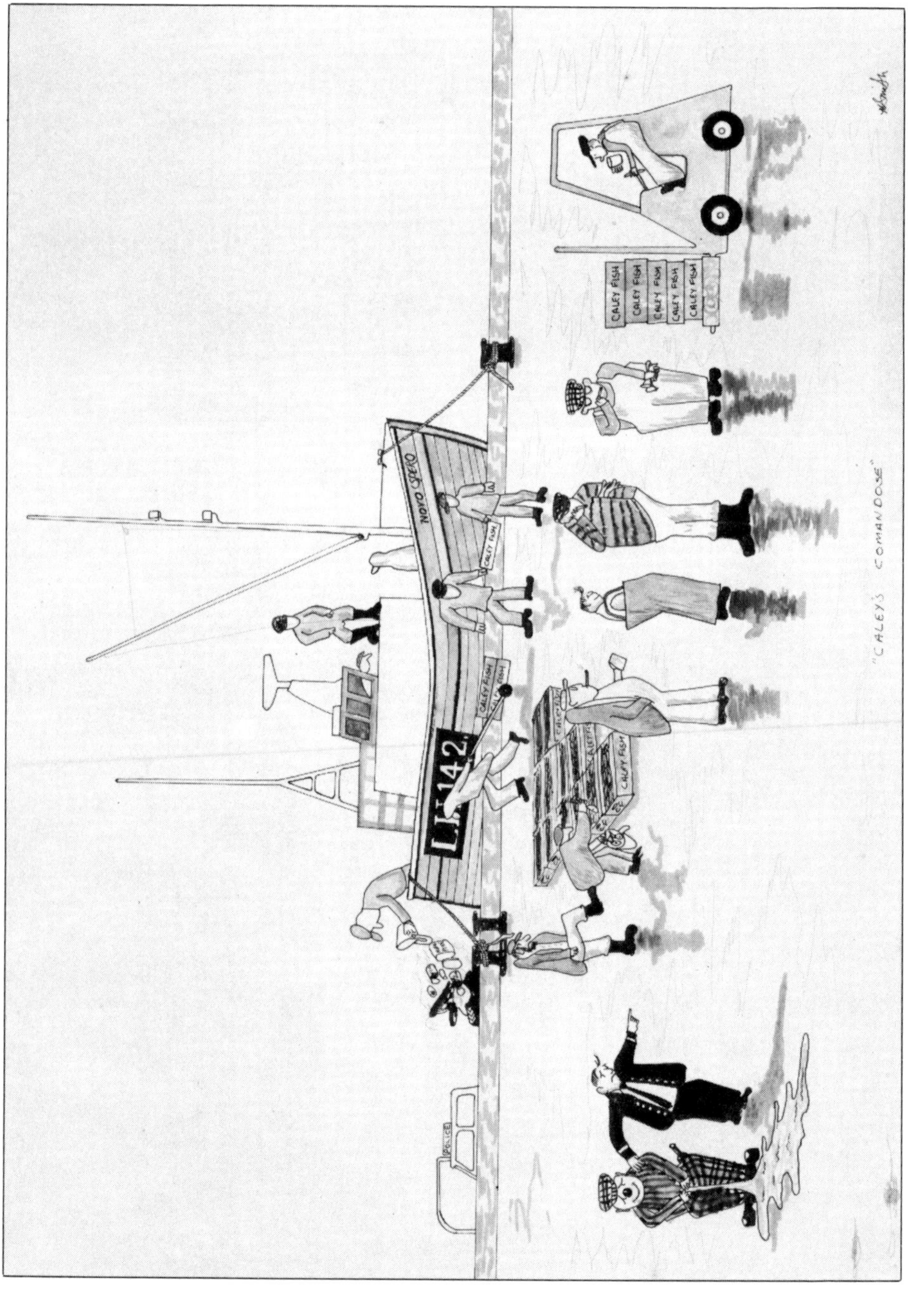
"CALEY'S COMANDOSE"
NOVO SPERO
LH 142
CALEY FISH
POLICE

CALEY'S COMMANDOSE

"Seine net boats and trawlers arrive at Shields fish quay and land their catches onto the market, mostly ready boxed. The men employed by the agents to lay the boxes out in orderly fashion are called 'lumpers', one team from Irvin's, the other from the second agent, Caley's.

Here we see Caley's squad on a typical night. George has the tally sheet and is willing his men into some kind of action. Nicky (or Mr. Tetley because he is rarely seen without a cup of tea through the night) is driving the fork lift truck, and trying to avoid other members of the team, Cyril, Joe and Bob, who as seen, are eager to get the job done and get off home.

Behind them in the 'Novospero', the crew of which is landing their catch amidships. At the stern the Marine Division Sergeant appears to be avoiding a possible breach of a Port of Tyne bye-law by preventing an over zealous cook from depositing litter in the port thereof.

Elsewhere on the quay the security man is issuing summary justice to a 'dipper' (a person who illegally fills white plastic bags from the fish boxes) while the quay master is asking a happy night reveller to leave by the main gate."

Shovelling prawns on the open deck. *Photo Keith Pattison*

SCAMPI IS KING

The Castle Dawn

Years ago, fishermen would throw back any prawns they took in their nets. Prawns were a poor man's food, they fetched no price. Most of them lived in peace buried snug in the mud at the sea bottom.

This has all changed. We are now voracious eaters of scampi (which legally should contain only prawn, though some establishments make do with monk fish). Prawn cocktail is de rigeur on every restaurant menu. We munch our way through tons of prawn tasti-snacks. The humble prawn is big business, and the fishermen know it.

There are some who would claim 'prawning' is not even fishing at all, who find something not quite right about digging the wee pink fellows out of the mud in their thousands, and hauling them aboard (where, incidentally, they suffer instant blindness). But when the hunting's good, you'll see up to half the Shields fleet 'on the prawn'. On one of my trips I counted 17 other vessels within view, all prawning. In stark contrast to the great loneliness of other trips, here was the sense of a giant boating lake, and also the seeds of suspicion that fishing at such intensity could see the prawn go the same way as the herring and the sprat — a vanishing act.

Ollie Allcock disagrees. Prawns he says are such prolific breeders that there's plenty for everybody. Time will tell.

Ollie's boat is the Castle Dawn, a 43 footer, originally a ring-netter. Like The Frem it is 50 years old, a fashionable (if peeling) shade of pillar box red, whose bare deck — with only two small winches — holds at its stern a small wheelhouse like a battered garden shed, an open, shallow-sided boat which again, to the landlubber looks frighteningly vulnerable.

There is though, says Ollie, no sea which frightens him with his vessel. They are words I repeat to myself several times when the stern sinks way below the water, when she rock 'n rolls to such a degree (literally) that the wheelhouse itself dips into the sea.

"She can take it" smiles Ollie, (and he smiles a lot). He is probably the most gregarious of skippers, an optimistic man whose cheerfulness belies his working-life style. Ollie lives in Cullercoats, is 36, claiming to look 56 "because I had a hard paper round at school", and as we set sail the first day from the Tyne, I find myself hoping that his bravado never reads as an epitaph, given the sea's patience, its talent in waiting as long as it takes to claim its victims.

The sea is a constant source of amazement. At one, unchanging and forever changing it is the most animate of elements. Time over I am astonished to find how in the space of a few minutes this peaceful and benevolent Dr. Jeckyll, lapping gently at the boat's side, can become the demonic and howling Mr. Hyde, keen to make matchwood of those venturing upon it.

I was to spend several days with the Castle Dawn. Like all prawners she is a day boat, returning to the Tyne each evening to land the catch which is bought by a wholesaler. Prawns sell for a fixed price, and are not at the mercy of the early morning auction, and overall the life may sound less arduous than that of the long-trip fishermen.

In some ways yes. But consider this. On our first morning we left the Tyne at 4am, and returned to port 16 hours later at 8pm. Ollie goes back home, drinks a few pints in the local, buys a Chinese take-away (not prawns presumably), grabs maybe three hours sleep, then wakes for another day's fishing. On this morning he told me he had worked 20 consecutive days in same fashion, week-ends included, and that the only break had been a broken clutch coupling keeping the boat in harbour for a day. Just a few days of this timetable brought home to me how much your life then belongs to the sea.

"You get used to it" says Ollie "and anyway, come January and February, we'll probably be unable to put to sea at all, which means no money, so we go when we can."

I bumped into Ollie in January, after my trips were finished. He was still grinning. Yes, he had been out that morning, desperately cold and hostile though it was, and his hands had stuck to the wheel. Yes, he'd be out the next day too (a Sunday).

Let's look on the bright side. Prawns don't smell. And there's none of that blood-soaked gutting on board. There's not a great deal to The Castle Dawn; the wheelhouse is a small cramped affair. There's a vertical exhaust reaching up through the roof. This becomes almost red-hot during the day, and you could almost tell the boat's crew from the tell-tale burn marks on their arms and hands. The automatic pilot is a piece of string tied to the wheel, the foc'sle is down a tight, six-step ladder — two bunks and a table, total height less than five feet meaning you walk around impersonating Quasimodo. Non-experienced seagoers such as myself at times scurry 'down below' to escape the elements above, only to find that its sense of claustrophobia, and stifling atmosphere brings on a queasiness, and you're forced back to the dubious delights of the fresh air above.

Ollie has a regular crew of one, 19 year old Mike Nugent, a 6ft 4ins man mountain who claims back home to have "eleven useless pieces of paper". They're nine CSEs and two GCEs which are little use to him out here. Mike agrees that fishing is a hard life, but says "what other lad of my age can earn £200 in a good week?"

When the Dawn was built, all hauling was done by hand; hence the shallow deck. Falling overboard here looks as easy as tumbling off a log.

Our destination is The Phantoms, North North-East of Tyneside. At 4am on an Autumn morning it's dark and cold. The moon is a finger-nail above the metallic sheen of the sea, the banks of the Tyne are shadowy, half-defined shapes.

There's only the soft chug of the 160hp engine. Ahead is a small twinkling flotilla, a diamond lit armada as the prawn boats head out while most people are still snug in their beds, and will be for many hours yet.

Ollie's been married twice, and has always been a seaman. Does he ever reflect on life ashore?

"Sometimes. I know I miss a lot, things are happening and I'm out here. It's inevitable when you're a fisherman."

At their stern, prawn trawlers carry the 'doors', rusty blocks of iron weighing two hundredweight apiece, slung each side like Roman gladiator shields. These are shot off with the net, and keep 80 yards ahead of it, ensuring it stays open. Someone is inventing the automatic prawn door, but at present they're unhitched and hitched by hand, a cumbersome affair. They can cost up to £500 apiece, and on one occasion a door dragged from the sea bottom suddenly spouted a thin spume of water, like a whale. It had been forced in under pressure, and shot out when same pressure was released.

There's no bobbin hauler on this boat. The net is shot by hand, and though the cod end is winched aboard, at the end of each day, the nets are pulled in by hand, a back-breaking task, especially precarious in a heavy swell.

We head out, in front and around us the diamonds twinkle in the formless dark, and when the sun does eventually drag itself up above the horizon, it drags with it a thick mist, visibility within seconds dropping to less than 50 yards.

Now we have a scary game of Blind Man's Buff. It's like a heavy white woollen blanket has been wrapped around our heads. We know there are boats all around us, and the radio in the wheelhouse buzzes with different skippers giving their locations. Our boat, like several small ones, has no radar, and other craft loom up frighteningly from the fog. The dangers are two fold; one that we may hit another fishing vessel, secondly that a bigger boat en route to the Tyne may plough into us.

"Those big ships, they don't slow down for the likes of us" says Ollie. "I remember once in the fog, I could hear the ship, hear the voices of the seamen on board, but couldn't see it. I just knew it was somewhere, and getting closer."

It's a strange non-world suddenly. The gulls have gone quiet, and disappeared, the sea's vast horizons have shrunk to a few yards. It's like being a blind man in traffic. Yet fishermen come out to fish, regardless, and that's what we do. The nets are shot and hauled in. We have one eye on the fog, one eye on the nets, one ear on the radio. The third eye (and fishermen always need three) is on the hauling chains, shot out along deck and given to sudden swishes, changes of direction which can catch you unawares (as

Ollie in his wheelhouse — "a battered garden shed". *Photo Keith Pattison*

elsewhere there's no protection from any machinery). Five years ago Ollie saw these chains knock a lad overboard. He drowned.

The North-East coast, as many bewildered holidaymakers have found is given to these sudden fogs and frets.

Up comes the first haul. Unlike the bigger boats, the skipper here is closely involved with the deck work, not a remote figure in the wheelhouse. The cod end is winched up from starboard, a great fat balloon of prawns. The knot's released, they spill out onto the open deck. It's a strange pink mountain of wriggling feet and antennae; they're shimmering, almost flourescent, more insect than fish, creatures suddenly pulled 250 feet to the air from their cold dark mud.

Some lurch across the deck, imitating the gavotte from Alice in Wonderland. others wave their front claws, like an orchestra conductor with two batons. For others there is a strange slow motion dance of death, while others still use their claws in some futile terminal contest against their fellow prawns, clamping hard on the other's body or limb. It may be hours before they die. Their ebony eyes look button-hard; their bodies are given to sudden convulsions, the coiled tail flicking back and forth. At times they resemble some bizarre clockwork toy inadvertently left running.

It seems ironic that these creatures, newly blinded should be sorted by we humans, temporarily blinded by fog. Yet the fog now lifts, and the sun shines through. The rays penetrate their very being, as it would shine through jelly.

The process is basic, primitive, Micky, Ollie, and the casual crew member Joe, shovel them up into plastic baskets similar to those in the launderette. From here they are sorted individually after being thrown onto an open tray precariously perched on top of two fish boxes. There are several sizes from the smallest, beetles, to the largest, clonkers, Ollie holds up one of the largest specimens.

"Some daft bugger will pay £20 for five of those" he says. Once sorted (and it's a long process, the prawns often taking a nip at the humans' plastic gloves), the prawns are washed by the 'donkey', a high pressure hose which uses sea water. Again there's an irony as they are temporarily put back in touch with the element that could save them — water. It drenches them, it transforms them. Where they were dull pink, clodded with mud, now they are glistening bright like freshly bathed babies. They look their best for their final moments.

On such a small boat as this, contact between crew members is much closer. Ollie and Mike have a relationship not uncommon among tiny crews. They regularly insult one another, swear, question parenthood and virility but not with true malice. At separate times Mike tells me he admires his skipper, and Ollie's faith in his bright crew member is such that come the Winter he'll hand over the boat to him for two weeks. Micky began on a YTS scheme, took City and Guilds in seamanship and says his dad skippers the Snowflake "but didn't want me on it". Why is he a fisherman?

"What else would I do?"

Joe is similar to much of the casual labour that hangs around Shields quay. He'll get a tenner a day for his efforts, and sometimes will sleep on the boat, while it's in dock, a sort of unofficial night watchman. He's also given to long sleeps while we're at sea, a fact which seems not to disturb Ollie at all.

The Dawn is a rusty old boat, but Ollie's proud of her, takes me down to look at the 3½ inch planking. "Compare that to a modern boat" he says "they're the ones that always sink."

All day Mickey slugs Coca-Cola from a giant plastic bottle. He has with him a ghetto blaster big enough to roast a turkey in, and he sorts prawns to the sounds of Elvis Costello. Like most skippers Ollie smokes too much, and on this boat he has time to. Shooting for prawns is less arduous. Each trawl is maybe five hours. Sorting washing and boxing the prawns (which are stored on deck — there's no ice or fish room on the boat) takes maybe two hours, then you sit and wait. Tedium is the enemy here, with deck quoits and dancing classes at a minimum. Down below, Joe cooks breakfast on a gas heater that doubles as a fire. The saucepan has no handle and water's in a plastic container. No toilet. The foc'sle rapidly becomes hot and stuffy from the gas fumes, and having such a precarious gas fire down here (you cook on the floor), surrounded by these old timbers seems none too wise. The gas cylinder is stored on deck and a rubber pipe runs up the ladder and through the hatch.

Somehow he cooks eggs, bacon and beans, and makes tea for four people using one makeshift 'ring'. Blue smoke pours up out the hatch, and the foc'sle is a fug of burning bacon. It's down to Joe to supply the skipper with tea in the wheelhouse. This can be a precarious journey up the ladder and across the open deck, and when there's a swell on, Ollie may have to be content with half a cup, the rest slopping out in transit.

"Sometimes" says Ollie, "I just seem stranded in here. One time I was in the wheelhouse for two days without a break, speaking to no-one, just me, Cigarettes and the radio. You think you'll go loopy."

He's a proud man. Come the winter, when fishermen can't fish, they're allowed the grand total of £5 a day from the DHSS. Ollie has never collected a penny in his life and doesn't intend to. One year he went ten continuous weeks without a penny's income.

"Whatever money comes to me, I like to earn" he says.

On this first day we returned to Shields with 16 boxes of prawns, one box of 'fish'. Half goes to the boat, then Ollie and Mickey share equal portions. Three trawls in a day, and it's debatable which is preferable, these long, five hour sessions, or the quicker working pace of the seine netters or the anchor bashers. Tedium as against exhaustion. Whichever, you're a prisoner of the boat for the duration. Two accidents mar the first day; firstly on the first haul the net rips, and many prawns escape. Back in harbour as the boxes are being winched 12 foot up from the deck (it's low tide), one load slips, and maybe £50 worth of prawns spill out into the dark cold water of Shields gut.

Prawns are best fished on the slack tide, and against the current. On this first day, when the mist lifted, the sea was becalmed, the warm sun coming out in a clear, almost windless sky. We worked in t-shirts, probably the gentlest conditions I would experience in the six months.

On the next trip the weather worsened, and the photographer who came with us was grey-faced throughout, occasionally tossing his head over the side as his stomach muscles rebelled yet again.

And on the third trip, a salutary experience, a small boat in a force eight gale. A few days previously Ollie tells me they'd been out in a force nine (storm force) and he'd been forced to clamber out onto the top of the wheelhouse to secure the loose life-raft. This I would guess is akin to climbing out onto a 'plane wing at 50,000 feet.

At force nine, the sea begins to 'smoke', the wind whipping off the surface into a fine spray, whether the waves are breaking or not. At Force eight you can sense this about to begin. We battened down the hatches, secured the foc'sle, crammed ourselves into the tiny wheelhouse. In this kind of sea, this kind of boat, trawling is impossible; the net is winched in. It's also impossible to sort prawns on an open deck in such conditions, so the bulging cod end is dropped, still secured onto the deck, spreading out like a badly made pudding. The wheelhouse leaks, there's insufficient room to move around, so as the waves crash over the top, slow trickles seek out heads and necks.

Being only eight miles from shore we can run for home. If you're far out in this weather, the best bet in a small boat is often to drop anchor, sit it out.

"I was once 180 miles out, in the Conmoran the weather got bad, and we just sat there for six days" says Ollie. "It was purgatory, you're helpless, you can do nothing but sit."

Running for home in high seas can be a long business. Top speed drops to two knots. Huge canyons open up before you, the boat first rears back, then drops down into them there to meet what feel like solid huge walls of water bringing it to a temporary shuddering halt before the poor engines can recover their breath and slowly force her forwards once more.

After canyons, you climb mountains, and there's that lurching stomach feeling, that sensation known to all travellers on the big-dipper, that now you're going up, soon you'll come down.

"You know" shouts Ollie "people pay a quid a time for this at Spanish City Funfair."

"Aye" Mickey yells back. "But the buggers can get off after two minutes."

The torrent of sea is against — bizarrely — a bright sun ahead, and a look behind discovers a permanent rainbow following the boat, a strange piece of beauty among the wild elements. The deck is now constantly awash, wave after wave flooding it, the stern dipping beneath the sea's surface like some feeding bird.

And the best thing is to stay silent, to quietly bite your lip and stare at the distant shore-line which seems never to grow closer. I *was* scared, and would have been more so without Ollie and Mickey. Their humour to some extent deflated the fluttering panics which accompanied the biggest waves, the most sickening lurches.

Fishermen live within close proximity to death and danger, they combat an element which at any minute could pick up their frail craft and smash it to tinder; consequently they treat it with an alarming, though at times reassuring indifferences. Just as the person who takes one 'plane journey will be more frightened than the stewardesses who do it daily (and therefore statistically are in much more danger) so it is with fishermen and laymen. Had I taken this return journey at the start of my six months, when the shade of green was even more pronounced, I may well have died of fright, instead of merely making a will.

Rarely have the Tyne's two welcoming piers embraced me so warmly. As we sail up the river to Shields, we check the foc'sle. There's two inches of water flooding the floor and Mickey's turkey-roasting ghetto-blaster is sopping wet.

The 16 hour day has turned into a eight hour one. I climb off the boat on shore. A quay side stroller looks down at our one cod end full of prawns.

"That all you got?" he asks. "I wouldn't go out in a sea like that, just for that amount." And he walks off, shaking his head. Who would go out in a sea like that, for that amount?

Only fishermen.

Hardly room to swing a cat — Ollie Allcock in the Castledawn wheelhouse. Photo Keith Pattison

THE MISSION

Why do fishermen, as against say joiners, accountants, or lorry drivers, have their own mission? Why should men whose behaviour on shore at times is more godless than most, need recourse to a religious establishment? You'll find the Royal National Mission to Deep Sea Fishermen in every main port.

Its name is an anachronism in Shields — there is now no deep sea fishing — for the rest it's an essential part of the quay landscape and culture, a sometimes spiritual sometimes material provider for the men of the sea.

And the answer to the above question, for this writer at least, is that sense of always being within breathing distance of death, that sense of your insignificance in the face of the natural elements concentrates the mind wonderfully (and often fearfully) and often puts it on the track of a possible greater force.

"Fishermen are more religious than we give them credit for" says Geoff Chandler, the 41 year old superintendent at the Shields Mission. "Yes, they're a tough breed of men, but religion is often a private matter with them. They tell me that at sea they often feel close to God. I understand this."

An articulate, intelligent man, Geoff Chandler confesses his is a unique relationship with the fishing community. He is not a reverend, but neither, he quips, is he irreverent.

"The last thing I want to do is preach to people. If I want to talk about a spiritual dimension in people's lives, then I have to earn that right."

"I want to meet them at their point of need. This might be either spiritual or material."

Which means they may come into the mission either to pray, or for pie, chips and peas, with many options between. There's accommodation for stranded fishermen, there are showers, a rest room, a chapel, a cafe which serves some of the cheapest food in Shields (and is now open to the public). I never saw the slightest trouble, or suspicion of it.

There's been a fishermen's mission in Shields since 1899 and the present one was opened in 1953 by the wondrously named Commander Kit Kat, a squat, unremarkable looking building on the waterfront. It's interdenominational, broadly evangelical, and free of dogma. Geoff Chandler is partly a manager, partly a pastor. His might sound a nice life but consider the task he

"Neither reverend nor irreverent"
Geoff Chandler of the Fisherman's Mission.

Photo Tony ***Patterson***

was once given of going to the families of seven men lost at sea on the same night to offer comfort.

When Tim Stone perished off the Tyne (see page ??) it was Geoff who was called upon to visit the widow and two young children. You might find him visiting retired fisherfolk, the old, the sick in hospitals. You might find him on the boats.

"In a way I'm like a social worker but without their restrictions. The fact that this mission is funded by public subscription, while not making it that secure financially, does give us a freer hand than officially sponsored bodies. I like to think that if a fisherman is in genuine need, whatever it might be, I can help without much red tape."

Geoff moved here with his wife and two young children in 1983. He's spent almost half his entire life working for the Mission, service in Fleetwood, Scarborough, Grimsby, Hull, Lossiemouth and Ayr.

It's one of those rare establishments that genuinely appears to have no axe to grind, no particular vested interest it jealously promotes, a charity in the real sense of the word, and though somewhere in the ranks there may be infighting and politicking, on the surface it seems refreshingly free.

How does Geoff Chandler see the Shields fisherman?

"There are" he says, and chooses his words carefully "basically two types. Those who are sea-goers, and those who aren't. Some men strain at the leash to be back out there, others, well, you see them round the quay, the weather looks perfectly fine, and you start thinking . . ."

As the Shields quay threatens to become a fashionable hunting ground for the yuppie and rising middle-class communities (new restaurants, discos, well-to-do flats and middle-class housing are springing up in the vicinity), landlubber trade at the Mission could grow. There's now a large blackboard menu displayed outside, and, though these new quayside additions are of little benefit to the fishermen themselves, Geoff Chandler is more optimistic about Shields future as a port that most.

"I know we've lost the deep sea trawling, but at least now there seems some co-operation between merchants and fishermen where once there was only conflict. They've realised they can't afford conflict. And they're beginning to market fish better, there's a steady build-up of visiting boats in the gut, and I think the dock dues are more favourable than say, Grimsby. Of course to get more boats you need buyers, and new buyers only come when there are more boats — Catch 22. Plus which the EEC batch licensing has hit the port hard, but as an outsider, I do see signs of hope, if that doesn't sound too optimistic."

Geoff's also a keen and talented amateur photographer, whose dramatic photos of La Morlaye, which ran aground in heavy seas at Whitley Bay in April 1985, were published in the Mission's national magazine. As befits this self-effacing, articulate, peacefully modest man, there were no credits given.

Inside the Fishermen's Mission, framed on a wall, is a poem written specially for the place by a group of visiting London schoolchildren. I think its simple lyrical quality makes it well worth putting in this book.

GEORDIE LULLABY

Every lad that's born in Shields
Has the river in his eyes
Hears the tides of Tyne turn slowly
In his heart until he dies

Every lass that's born in Shields
Runs the river through her heart
Knowing 'fore she's one and twenty
Love and lover seaward part

Every bairn that's born in Shields
Urchins of the riverside
Golden lads and lasses ever
Swept away upon life's tide

WITH LOVE FROM JUNIPER HILL, JUNE 1985

JIM ROBERTSON AND THE SMALL REVOLUTION

Fishermen are a conservative bunch, and the fate of Shields men who invent things is to be laughed at. It happened to Purdie with his steam trawler and it happened to Jim Robertson with his rock hopper.

Jim's a Shetlander, a former master mariner in the Merchant Navy who worked for the White Fish Authority in Hull, and in the late 70's set up his own net making firm on the fish quay, a redbrick unspectacular building, but the clue lies in the pile of old car and lorry tyres outside. A Shetland skipper asked him if he could make a new type of net to catch cod (at that time fishermen were knee deep in herring and whiting, and landing their quotas in a couple of days or so).

"I'd already had this idea, and I suppose that pushed me into it. The skipper was Jimmy Fullerton of the 75 foot Don Vale, so I made him this net and took it up to him. The crew laughed at me, and said it would never catch fish. I ended up putting it aboard myself. On the first day the boat landed 240 boxes and never looked back. It was outfishing the big trawlers from Aberdeen. Now everyone knows the rock hopper, and they even use it in the USA."

"So what is it? Before this they used heavy metal bobbins to pull nets along the sea floor. They were awkward, could only tow a certain amount and got caught up on the rocks. My idea is simple, a series of circles which bounce or hop over all obstacles. They've been a great success."

(And in the week I went to see Jimmy, Fishing News was reporting an £86,000 record catch of squid north of the border using his invention).

"The rock hopper takes a net's headline from 78 to 110ft, the foot rope from 130 to 190ft. Before this a net stood up six feet, now it's 40 feet high."

"There's no such thing as a patent with these things, so yes there are several people imitating them. The difference is I design my own special nets, though exactly how is my secret."

"Made a fortune from it? Well, put it this way, without the rock hopper I'd have gone out of business."

I asked him about fish stocks. Some people claim his invention pillages the ocean floor.

"So what are we supposed to do? Go back to rod and line? My father said seine netting would finish fishing. It didn't. Cod could go the way of herring, but the herring came back didn't they?"

Jim Robertson in a sea of nets

Photo Tony Patterson

"I don't really think I'm helping overfish the sea. We British tend to abide by the rules, even if others don't."

Jim Robertson's rock hopper typifies fishing's dilemma. How to advance technologically without damaging the sea's wealth. Everyone wants to catch fish more efficiently, cheaper, quicker. But what about the long term?

The giant in his castle — Congener skipper Allan Morse (6ft. 5ins.) in the boat's wheelhouse. *Photo Tony Patterson*

UNITED WE SAIL

The Congener

Brass bands and ticker-tape may not figure a great deal when a fishing boat sails from the Tyne, but for the larger vessels at least there is some small sense of an expedition setting forth. The 80ft 250 ton stern trawler The Congener is the second biggest boat in the fleet. Like many of the best predators she hunts in pairs, and the sight of The Congener without her sister ship the Camellia would be as rare a sight as Laurel without Hardy.

Too big for the shallow draught of the gut, the pair trawlers moor up 100 yards upriver at West Quay, and on a breezy September night a small cluster of wives, sweethearts and friends gather as the nets and chains are spread out, measured and wound in, stores loaded. I'm never sure whether fishermen come to terms with sailing from the Tyne for a long trip (ours will be six days). Certainly I don't, and those last views of the jetties, the castle, the priory, that sense of the sea swallowing them as we head off into the night brings the normal mix of sadness, panic and appreciation.

Someone once said of being at sea that it had all the disadvantages of being in prison, except that with the latter there was no fear of drowning. It's not an aphorism I'd repeat at sea itself, and with the Congener there *are* definite advantages over my previous trips. In many ways this ten year old vessel is the best equipped in Shields. The skipper and crew of six (including two trainees) are offered such luxuries as reading lights in the bunks, a separate mess room and galley and — the first time for this novice seamen — a toilet. There's also a sink, a shelter deck, a whaleback (the protective covering over the bow), and though the pillows still feel damp — one crew member discovered mushrooms growing under his mattress, and the living accommodation would scarce impress the idle rich sailors at the Boat Show, it's more than a cut above the frugalities of The Frem.

One reassuring aspect of pair trawling is that of always having another vessel close by, a small lump of security for a seagoer not yet sufficiently seasoned to be blase about the chances of sinking. The boats work in harmony, taking turns to shoot their 150 foot headline net trawled between the two from the side, 400 yards of rope and chain on the sea bed. Each boat will shoot twice a day, four hauls with as little as five, or as many as 120 boxes coming aboard per haul, mainly cod and codling, but a smattering of sole, witch, cat and dog fish.

"Every time we leave The Tyne, I ask myself, what the hell are we doing here." This is Austin Houghton, Ossie to the crew, deckie and chef, curly-haired, bright-eyed, sharp-witted.

Ossie's sense of humour proves an important antidote to the tedium and depression long journeys can bring, and on this trip, despite some heavy swells which make cooking as much an exercise in juggling as gastronomics, he'll serve up for us chicken curries, chilli con carne, roast beef and yorkshires, joints of lamb, casseroles, gammon steaks, beef steaks. Food isn't always a welcome sight when you're in a frisky sea, but starving yourself rarely helps combatting a queasy stomach.

In years gone by, when fishermen had to fund their own food at sea (now the boat provides), fish was the staple diet. But you see enough of the damned stuff day in, day out, and with the exception of a rock turbet cooked in white sauce, this is definitiely a meat-eating trip.

Ossie battens down his saucepans with strong twine, ties up the 'fridge door. Why did he become a cook?

"I was volunteered" he says, whisking up some batter. His galley is small for a one-bedroomed flat, large for a fishing boat. He rather enjoys cooking but would never dream of admitting it.

The Congener has three levels. Down below, forrard is the cabin, eight bunks each with its own privacy curtain. Like most cabins it's triangular shaped, small, no natural light, no tables, chairs, nothing but sleeping space and a small patch of floor. Midships is the fish room (the largest space on the boat and an ironic reminder that though the fish are in no position to appreciate it, they're given much more spacious accommodation than the humans), and the engine room aft. The middle-section has a half-covered deck, galley, mess room, toilet, small wash-room with non-functioning shower, and store-room.

On top is the wheel-house and a series of catwalks leading from it. The wheelhouse is well endowed with modern technology. You could say the boat is over endowed, or as skipper Allan Morse puts it "this is a Rolls-Royce doing the job of a taxi."

His dad, Allan Snr., who once retired from the sea, but had to come back, skippers the sister ship The Camellia. Fishing's often a family affair; cousin David is on The Camellia and another cousin Norrie is here with the Congener.

Allan Jnr is a man-mountain. He stands 6ft 7ins tall, weighs around 19 stone, and (when he can find a pair to fit) wears size 14 shoes. The big black beard and shock of black hair, deep voice and piercing eyes give an impression somewhere between Geoff Capes and Bluto, and overall you feel rather like Jack when he met the big lad at the top of the beanstalk. I'm scared stiff of him, if only because of his enormous presence, also the knowledge that here in the wheelhouse is most definitely Allan's territory. He spends most of his time here, food and drink are brought up to him by crew members occasionally summoned in booming terms. Alright, he wouldn't crunch your bones, and the trip also reveals an affable size, but first impressions are daunting.

Like most vessels, on The Congener every crew member is a 'deckie'; they all help shoot and haul, they all gut fish. But each of the four senior crew has a special responsibility. Ossie is the chef, Norrie Morse is in charge of the fishroom, for Mick Boult (who's been with the boat since she was launched) it's the engine room. Geoff Davidson is in charge of deck and nets.

If your car breaks down on the A1 you pull into the hard shoulder and curse while awaiting the AA man. At sea it can be more serious, a boat without an engine is out of control, at the mercy of the elements. On this trip, the Congener's diesel engine will perform for 140 hours non-stop (imagine that in a Cortina), and every hour during the day Mick Boult dons the ear mufflers necessary to combat the huge noise, and descends in to the roaring heat of the engine room. He's the quiet crew member, deep ponds of eyes and a black thick beard. Later I was to learn that Mick once spent 20 days out on an anchor basher, a spell which would reduce a less stable person to gibbering helplessness.

"Small things that go wrong, I fix myself" he says "she doesn't take much really. And if you do break down seriously — well, you send for help, sit it out."

We steam overnight, 70 miles East to Dab's Hole, awaking the first morning to a heavy swell. There's someone on board even more of a land-lubber than myself, Noel Patterson, friend of the skipper, curious to experience life at sea. The previous evening he was in a state of great excitement, this morning he offers up his stomach to the fishes and lies groaning in his bunk, face the colour of a dirty dishcloth. For two days he'll suffer that unique sense of seasickness. One trainee, 16 year old Tony Noble has also staggered deathlike to his bunk; the other, 17 year old Peter Ingram, who unwisely downed a jar of pickled whelks before we sailed, will spend the next day in a similar agonised state. Not to be outdone I vomit over the side. We've been at sea less than 12 hours. Four casualties.

Alan Morse tells me "the last visitor we had on board started puking as soon as we got away from the Tyne and he was in his bunk for nine days ill. He couldn't move. We hardly saw him till we got back on shore."

Even hardy fishermen get seasick at times. This was the worst start we could have had, an incessant heavy rolling swell leaving you queasy, exhausted. Getting through any day in such conditions is difficult, but here we have to fish. The chains clank out from the winding gear, the net is shot, the Camellia comes close to pick up its rope end. Eight times a day the two boats come within kissing distance then separate, putting a quarter of a mile of sea between them. When you're far out at sea, 100 fathoms below you, no sight of another vessel for days, this sudden intimacy is bizarre. You stare across at the other vessel bobbing a few feet away, rather like bus passengers stare in fascination at the interior of another bus. Nor are any words exchanged between the two crews at this time, there's hardly an instant of eye contact. Gear is uncoupled, cast off, wound in, shot, the business is done, the brief intimacy is ended, silently the sea between them grows once again.

It takes the sight of the sister ship to appreciate fully the movements of small craft in such a swell. We watch the Camellia rear up like a startled horse, dip like a diving bird. She sways from side to side like those fairground moonrocket rides. Even the slightest push it seems would see her roll over, yet always she rolls back, a few seconds stability, then rolling the other way, rearing, dipping. At times she is swallowed entirely by a great hunched back sea, reappearing a few seconds later in a deep water valley.

This is not good weather for either fishing or the human stomach, and our first hauls are disappointingly small, some under ten boxes. And again I'm knocked back by the sense of fatigue, partly the body's constant readjustment to the boat's movement, partly the intake of such pure unsullied air by lungs conditioned to normal industrial pollution.

We are aliens in this environment, a fact reinforced by the seemingly effortless bobbing of the birds, kittiwakes, petrels, the beautifully concorde-sleek gannets, who, it is said, die of starvation when their eyesight, blasted by the ice-cold impact of countless sea dives, eventually fades and they can no longer spot their target.

"Look" says Allan Morse, pointing out an arctic skewer. "They're the most vicious birds there are. I've seen them land on a kittiwake and steal the food from half-way down their throats."

The screeching blizzard of sea birds wheel and circle waiting for the fish guts to be tossed overboard. This may be an unnatural link in the normal food chain, but they don't complain.

There's some fancy gear on The Congener. Part of the shelter deck is operated hydraulically to swing open, enabling both easier shooting of the net, also emptying the cod end into the fish pound. In the wheelhouse, the electronic fish finder is sensitive enough to locate a single fish, or even a piece of rope at the sea bottom. The only problem is getting exactly above it. Every skipper has his charts (though computer tapes are slowly taking over). To the layman they are a series of scrawls, blue, red and green ink, pencil, criss-crosses, lines, dots. To the skipper they're jealously guarded records of past trawls, likely fertile grounds, places to avoid where you could get gear trapped, sand-banks, stretches of rock (better for fishing), dips, rises, lumps of old wrecks. When any skipper sets to sea, he's armed with minutely detailed information to help his hunting. To the layman it's a vast expanse of water out there, one mile the same as any other. To the fisherman it's as changing and variable as any urban landscape.

"Every year" says Allan Morse "the fishing gets worse. A few years back we could get 200 boxes a day in this area" (a box in Shields is seven stones — other ports have different measurements).

"Now look at it." Allan Morse blames several things, including the vast overfishing of codling round Heligoland a few years ago. "They took 100,000 young fish. They would have been half a million fully grown fish by now. People just didn't think."

"It was like the gold rush, no regulations, now we're all suffering. Why didn't the EEC do something?"

Each trawl is maybe nine miles, at a speed of two knots. Travel too fast and you distort the shape and angle of the nets and gear. Too slow and the fish outrun the nets. They can sense the build-up of pressure from the coming net. A fish says Allan can swim forever at its natural speed; above that speed they burn themselves out, which is why many come aboard exhausted, which is why their small red hearts can be extracted from their bodies, beating on in isolation for several moments like some ghoulish 'prop' from a Hammer horror movie.

The Congener was built at the height of the sprats boom, a time says Allan (and this is the closest he'll get to nostalgia) when boats would steam back into Shields so loaded down with sprats as to have virtually no freeboard (the amount of boat above the water line). He remembers limping back, in a gale, with only one inch of freeboard, more than 100 tons of sprats on board. Built for sprats, the Congener was soon robbed of its natural prey. After years of being so thick in the water as often to ensnare the propellers, when vast shoals gathered so close to shore that the fishermen would chat to those on the beach, the humble sprat called it a day, and disappeared. Who could blame it? Two years after The Congener was built, in 1979, they vanished and never came back. The party was over and the Shields fleet couldn't believe it. The sprat, like the herring, left them punch-drunk. No-one's exactly sure whether they just swam away, were overfished or what. Nor if they'll ever return. Meantime the Congener plies the seas, a luxury, too fast, too big, too powerful for the job it has to do.

Allan Morse would like to sell it, but these days, where are the takers?

Day two, and we've steamed far north up to Devil's Hole, on a longtitude with the Firth of Forth. Rising on a fishing boat is unceremonious, a loud shout of "Get up!" even before the first rays of dawn are over the horizon. Tony is still groaning in his bunk. I lie in my bunk, drained, snatching a few extra moments, half guilt, half relief at not having to go onto that cold hostile deck quite as promptly as the rest.

But there are big fish to be had here, 20, 30 pounders, great grey battleships of fish. When these are gutted, the deck literally runs with blood, the foamy dark tides swish the intestines and guts from one side to the other, a horrible thick soup sloshing ankle deep round your boots.

On some boats I gutted for most of the trip. On The Congener I still lacked the confidence to muscle in, made do with deck labouring (washing boxes, hosing down the deck) and took up the spud knife in the galley.

Fishermen gut a fish with the speed most of us unzip a pair of trousers, a quick downward stroke of the knife and the poor creature is slit from gill almost to tail. There's another quick twist and out come the guts and intestines. Try this yourself and you're left yanking and slashing while what resembles bloody spaghetti hangs half-severed from the gashed belly. It stinks too.

On the second morning the crew need all their gutting skills at the promise of a big haul. You can sense the excitement, the anticipation, and as the net is winched in, there's a silver flash in the water, a long silver tail pinpointing the cod end (the final bulbous part of the net into which the fish are driven). This flash of silver is always a good sign and whereas on an average haul the gilson will lift the entire cod end in one operation, this time it takes six lifts, fish queuing up the length of the net as the cod end is filled, emptied lowered, and filled again from the net. Too heavy a lift and your boat can capsize.

In this one haul — taken in not by us but the Camellia (proceeds are shared equally) — are 135 boxes of fish, a £7,000 bonanza. It's such a big haul that both ships share the gutting (normal rule is you gut your own haul).

The sun is out, the sky has cleared to a fine blue, the swell has dropped and suddenly the entire mood has lifted. The fish are standing in line to come aboard. Add to that Ossie's midday meal, roast beef and yorkshires, followed by apple pie and custard.

"It's big, but I've seen bigger" says Geoff Davidson. "Three years ago in Heligoland we took 600 boxes in one haul. We were on deck gutting non-stop for 49 hours."

The fish is gutted and washed on deck, thrown down a chute to the fish-room, where it's boxed and iced, wedged tight in case of bad weather. The fish room's filling, but it's hard to imagine it full. It's the size of a medium restaurant, and though there's no actual refrigeration it still looks like the hall of the ice-king with the 15 ton of powdered ice taken on board in Shields. Norrie Morse is the man in charge down here, a cold dark cavern with no natural light, no natural entrance (you clamber down the chute). Ice doesn't stay powdered long, solidifying into great white walls which Norrie attacks with a pick-axe, resembling some arctic explorer. This ice will stay unmelted incredibly for 11 days or more. Each box is lined top and bottom with ice and the stacks slowly grow. It's a strange twilight world down here, and a fisherman's waterproofs seem scant protection against the cold. These waterproofs incidentally are not provided. Each fisherman buys his own, plus boots, gloves, an outlay normally of more than £50.

The second night out, 8.30pm, fishing done for the day and Allan Morse descends from his wheelhouse for some recreation. It is one of the most remarkable metamorphoses I have seen. A tattered piece of baize is spread across the messroom table, cards are produced, margarine tubs are positioned by each man. The game's on.

And the man who in the wheelhouse seems a giant troll, a remote and awesome figure whose authority brooks no question is suddenly a laughing, joking, easy-going card-player, wisecracking with the crew like any lad out with his mates. Even allowing for the required daytime discipline, and more relaxed atmosphere of the evening, the transformation is astonishing; it's as if he actually *looks* different, eyes softer, features less formidable. The change is so convincing that one such as myself, climbing the next morning to the wheelhouse is pulled up short to realise that the Jeckyll and Hyde formula has now been reversed again, and we have Allan Morse, skipper.

Fishermen swear more than any people I know. A game of cards brings a non-stop succession of insults often of the most revolting nature, doubts brought to bear on the others' parentage, sexual leanings and potential, honour, family, fetishes and personal habits. Were these nice polite middle-class folk, such insults would probably lead to writs, or actions in the courts. Here they are bandied about naturally, and I never heard an insult so bad as to provoke true hostility. So where lie good or bad manners?

Interesting to spot them playing cards; Ossie, small, wiry, Geoff, deep chested strength, Norrie, a cool dude at cards, Mick, reserved a loner, usually opting for a book in his bunk. Apprentice Tony stoically suffers what every fishing apprentice suffers, perpetual humiliation. This is the only boat with such evening pastimes. Pair-trawling is not as consuming as some types of fishing where the men slump exhausted into their bunks.

Day three and the skipper is sounding forth about the poor plight of Shields as a fishing port.

"Fifteen years ago they planned to build a whole new quay. If they had Shields would still be attracting the big Scottish boats. But look at it, the facilities are archaic, the place is rife with vandalism. The Port of Tyne Authority (which took control of the quay in 1968) hasn't the slightest interest in the fishing fleet. They'd just as soon let it die away. The Humber fleets have been looking for years for a new venue with their crippling landing charges — they could have come here if we'd modernised it. The Tyne has been dying for 20 years and fishing was the only chance to revive it."

"But they've thrown the chance away. They're talking of a £12m development at Eyemouth — that's just a bloody hole in the shoreline."

"Fishermen have had to fight tooth and nail for every tiny thing in Shields. The only real improvement recently has been the ice factory and we built that with our own money. The PTA never give any help. This boat's moored at West Quay, and I never know when it will be vandalised, there's virtually no security at all. Look at "The Frem" — had the wheelhouse stripped nearly bare. Course I'm bitter. It's a disgrace. Basically the Port of Tyne Authority does not want a fleet in Shields at all."

By Friday the Devil's Hole has offered up whatever riches it is willing to. We decide to head down south to Dogger Bank, fish on a longtitude with Scarborough, and we steam through the night. Some boats drop anchor at night. These two never do, following one another like puppy dogs through the whole trip. We've seen no other boat for more than two days, and fish and birds apart, our only other sign of life has been a distant pumping station.

In the messroom the initial curiosity at the girly mags has subsided. They lie discarded, scarcely looked at (a moral here Mrs. Whitehouse?). As on other trips, I'm surprised how well fishermen react psychologically to their strange environment, a semi-solitude yet living hand-in-glove with other men not of your choosing. There are grumbles of course, behind-the-back comments, petty feuds, but at the base all seem to realise that compatability is a necessity. They are, literally, all in the same boat. During a lull I ask 16 year old Tony, will he see it through, will he become a fisherman.
"Yes" he replied, "I like it" and he insists on the answer, even in the face of

being treated as the working class equivalent to an Eton fag. People bawl at him, he makes tea, washes up, cleans tables, he's the dogsbody, unable to answer back. On one occasion Allan Morse bellows so loudly at the poor lad I felt sure his limbs would drop off in fright. He grins sheepishly. "I *do* like it."

Geoff Davidson is less enthusiastic. A thick-set tattooed man of deep guttural tones, he warns Tony: "Get out while you can. I hate the life. I want to work on the rigs. The only thing in favour of working here is that it's easier than some boats."

It is too. With only two hauls per boat per day, even the gutting and washing leaves long stretches of inactivity. Some crew grab quick kips in the mess-room, others read the cheap novels with which the boat is well endowed. But it's not only the working hours you have to contend with. Unlike normal jobs, off-duty there's simply nowhere to go. The factory worker, shop and office workers, they close the door behind them, go home, change, watch telly, go to the pub, put it all behind them. On a small fishing boat this is impossible, it is there constantly. The smell, the closeness, the people, the sea's movement. You cannot escape. At times this matters not overmuch; other times the trapped feeling is overwhelming, a terribly frustrated desire to run, jump and be free of this small bobbing container. There is no radio (the wheelhouse apart), no TV, no video, no newspaper, no stereo.

The mess room is not oversmall (smaller than your average house bedroom). Yet like the cabin it has no natural light. Often the worst aspect of fishing (and several take prominence depending on the circumstances of the time!) is the impossibility of even temporary escape from that huge flat ocean, a soulless horizon spinning round 360 degrees.

"Get this down you!" chirps Ossie. It's his meal of rock turbet (otherwise known as dog fish) in a tasty white sauce, oodles of it, and cooked with such skill as almost make us forget it is *fish.* "You've never eaten any fresher" he says, and it's true.

Trawling a large net, unseen between two boats for nine miles along an unknown sea bed throws up problems, the chains, the ropes, the tides, speed, the angles, these are all factors to be judged. Even the most experienced fishermen couldn't anticipate what happens on the Saturday. The net snags, the boat manoeuvres to free it (a snagged net not only risks costly damage, but offers an escape route to the fish already caught) and eventually it comes away. When they winch it in it's to find the entire headline has been sheered clean off.

Allan Morse runs it through his fingers. "I've never seen anything like it" he says. It's as if someone had taken a giant razor blade along the net's whole length. Think of someone unseaming a pair of trousers, severing the cotton strand by strand, leaving the material untouched. That's the effect.

Now this is a sewing job and a half. There's a full 140 feet to do by hand, using thick twine. It takes three hours, an exasperatingly slow task which merely puts us back to square one, and meantime those little fishes are all swimming away down there as cheeky as you like. At such moments fishermen, often criticised for loud, excessive behaviour are amazingly stoical, they set about the work, inch by inch without a word of complaint, the giant 'needles' weaving in and out until the job is done, the net is reshot.

By Sunday there is a sense of the trip winding down. How many times do we look across at The Camellia, so near, so far, and wonder at their trip. We see them for 24 hours a day. We know nothing of them. We've moved again so that now we're directly East of Tyneside and on the day's final haul the boats steam towards a huge molten evening sun for a 4am landing. Between us we've taken 425 boxes of fish and at Monday morning's auction this should bring in around £20,000. For a full crew member this will mean a wage packet of around £700. Those who think that's easy money for six days work are those who know nothing of the sea. It is not earned cheaply.

My own sense when I step ashore at Shields is one of fatigued relief, a small celebration that life can return to some semblance of normality, that those seemingly insignificant pleasures can once again be realised. A walk to the corner shop, a pint in the pub, a real bed, a bath, firm ground beneath your feet.

For the crew of the two boats, there are several hours of landing fish to come, the market, then home around 8am. When will they sail again?

They answer, unbelievably, the same evening, 8pm. The thought for me is almost unbearable, the need to take some relief from such a strange way of life so overwhelming, that the concept of such a quick return to that vast lonely ocean and that cramped life is one I can't even contemplate.

They see it this way. "The weather's good. We had a lay-off recently. Now we have to fish as much as we can." And back to the sea they go. A few hours later.

NICK HEDGES

The photographer Nick Hedges was commissioned in 1979 by Side Gallery Newcastle to document the North-East fishing industry, and he chose to concentrate his work on North Shields. Nick spent three months taking photographs, the catching, landing, auctioning, and eventual retail selling of fish, and a further two months preparing his exhibition. More than 70 photographs toured libraries, pubs and schools in the region during 1980, and six examples are on the following three pages.

Phone 71684 74393
Joe APPLEBY Ltd
Shops Supplied in these Areas
DELIVERY SERVICES
Deliveries Daily
Grams 'APPLE'
ROUTE 1
BIRTLEY
CHESTER-LE-STREET
DURHAM
BRANDON
WILLINGTON
CROOK
WITTON PARK
& DISTRICTS
ROUTE 2
EASINGTON
HORDEN
SHOTTON
WHEATLEY HILL
THORNLEY
HASWELL
SHERBURN
& DISTRICTS
ROUTE 3
WREKENTON
WASHINGTON
HOUGHTON-LE-SPRING
HETTON
FENCEHOUSES
RAINTON
& DISTRICTS
ROUTE 4
SOUTH SHIELDS
HEBBURN
JARROW
SUNDERLAND
RYHOPE
SEAHAM HARBOUR
MURTON
& DISTRICTS
ROUTE 5
WALLSEND
WALKER
NEWCASTLE
GATESHEAD
DUNSTON
& DISTRICTS
ROUTE 6
FOREST HALL
THROCKLEY
NEWBURN
WESTERHOPE
GOSFORTH
& DISTRICTS
ROUTE 7
EXPRESS
SPENNYMOOR
SHILDON
COUNDON
BISHOP AUCKLAND
FERRYHILL
& DISTRICTS

Quieter days — retired fisherman John "Jumbo" Henderson. *Photo Tony Patterson*

THE RETIRED FISHERMEN

They meet every Monday afternoon in a quiet upstairs room at the Fishermen's Mission on the quayside. There's about a dozen of them ranging in age from the mid-sixties to the late seventies. There's the soft clink of dominoes, the faint slap of cards being laid down on the formica table top, the slight clatter of the trolley as it brings them tea and biscuits. Through the window they can see the various boats entering and leaving the Tyne. Age has probably mellowed them but their memories, as happens with the elderly, grow more vivid. Talking to them you sense again that ambivalence of fishermen; the recounting of the seemingly horrific conditions along with the insistence that it was "their life". These are some of the words of the Shields retired fishermen, most of whom had left the sea by the 1960's. They contribute 20p a week towards their annual trip.

"We were on the coal-burning boats, steam trawlers. All gone now of course. You'd take 100 ton of coal on board and use maybe ten ton a day. We'd trawl 24 hours a day and when there was plenty fish you'd almost need matchsticks to stay awake. We wore an oilskin frock, a sowester and a sweat rag. When diesel engines came along, it was pure delight, just press the buttons."

"If you worked in the engine room it was five hours on, five hours off for ten days. You were firing (shovelling coal) all the time except when you were raking out the ashes and oiling the engine every 20 minutes. Sounds bad but it was better than working on deck. You'd get regular sleep, you got a wash at the end of your shift, and you got a change of clothes. Not like those on deck who never knew when they'd work."

"Ships would have a crew of nine and in the early boats there was no electric light. We'd have oil, gas or carbide lights and they all had to be maintained. The North Sea was fished out before World War II. The war was a blessing for it, it allowed stocks to get back to normal. I suppose it was hard, but not for us. It was our job, it was all we knew. You'd be tired, you'd moan, but you had a comradeship you could never get in a factory, and you lived closer than any family could."

"Some of the crew's bunks would be in the foc'sle (forward). Crossing the deck was highly dangerous for them in bad weather. They'd need to put on oilskins and be hauled along on a rope to get to the wheelhouse."

"We had no gloves in those days — anyway gloves slowed down your gutting. Your hands would develop hard callouses which you'd cut off with a knife. Or you'd get salt water boils on your wrists where the material chafed. These were really painful. Your hands got so insensitive in cold weather you just couldn't pick anything up."

"Oh yes, we'd probably all be fishermen again given the chance. Before the war you got 8/6d per day (42p) and your food had to come out of that. You also got £6 out of every £1,000 earned. No wonder we ate so much fish at sea — it was free. Or you could go to South Shields college, try for your mates or skippers ticket. You could spend six months there with no income, pay for your own upkeep, books etc., then you might get a ticket, but no boat. In some respects skippers were worse off than the crew, because they got nothing if there was no fish, whereas we did at least get a basic wage."

"I remember coming back from an 18 day trip to the Faroes, and we didn't even have enough fish to clear the expenses. It meant I'd earned nothing all that time. I was a mate, and mates worked on a similar principle to the skippers."

"Fish weren't boxed at sea then, they were laid out individually in bays. It could take a long time to land and the tradition was they were always laid out facing East (towards the sea). Layers-out could tell you almost exactly the weight of a bay (five rows of ten fish) just by looking."

"In those days there were 47 pubs round Shields quay; in the 30's up to 250 drifters came into Shields every day, and the market was 24 hours a day, boats coming from all sorts of countries. Look at the prawns they're landing now — we'd have thrown them back, they weren't good enough. Mind, no-one ate prawns like they do today."

"Sometimes you'd sleep standing up on deck, waiting for the net. You got used to it. Now I'm retired I often can't get to sleep at all. I remember going on my first boat, the wheel was bigger than me — six feet in diameter."

"Fishermen retire when their hearts are broken. There's no set time."

"Remember that Dogger Bank sauce? Mustard and vinegar with a touch of sugar. Pour it over your fish. Lovely — and simple. I think they call it hessle sauce in Hull."

"I was a ships husband. My job was to go round and get the crew in. Sometimes I'd look in 38 pubs for one man. Often you found them, they were drunk and wouldn't come, so you'd put them under arrest and they'd be charged the next morning. I had a job to do and I did it. Often they didn't want to go back to sea, you see? But they had to."

"The old herring picklers bandaged each finger individually. They called it 'gibbing', pull out the gill and gut at the same time, so fast you couldn't see it. Once they started, and their head went in that barrel it never came out till it was full. Brined herring would keep for donkey's years!"

"The picklers would sleep up in Dockray Square, six to a bed, three at each end. I remember 'The Clog Vaults', it used to take a dozen in."

"I was on the St. Lawrence 90ft steam trawler during the war. There's a photo of her and you can see how we built concrete slabs round the wheelhouse because of the German bombers. We were attacked by bombers in 1940 near Longstone Lighthouse. The bombs landed half a mile away and they machine gunned us too. We kept rifles on board to fire back."

One of Paul Rogers' etchings from his experiences on North Shields fishing boats

Harry Horn prepares some more (non-Shields) herring for the ovens. Photo Tony Patterson

CURED FOR LIFE

Harry Horn's father cured fish in Shields, his grandfather, and his great grandfather; with his three sons in the business it seems unlikely the firm will pass into strangers hands. There's an irony about Harry's premises. They overlook the fish quay and the daily landings, yet the herring Harry buys to smoke into kippers are not landed in Shields, but Scotland, then transported by road. Ask Harry to show you his beautiful old wooden kilns. You can stand up in them, and the smoking process is little changed in decades; a pile of wood shavings on the floor, a match, then close the door, and the herring, hung like coats in a wardrobe, are slowly smoked. A pall of blue smoke rising from Harry's distinctive chimneys is one of the quay's landmarks.

"It's a pity there are no herring fishermen in Shields now. To think there used to be the famous 'Shields Kipper'. But since the herring ban, no-one seems interested in landing them in this port anymore, even though the ban is now lifted. They don't think the market's there. Buy a kipper on Tyneside and there is a 75 per cent chance it will be one of ours. We buy from Scotland but also the Manx herring. This is the way they've been smoked for years. Technology has arrived, and I was shown an American invention called Liquid Smoke. It looks like soy sauce, and a small bottle costs around £28. The herring are dipped into the liquid, but I reckoned it would add £1 a stone to the price of fish. It's not for us."

"Things are different for fishermen now. The philosophy seems to be 'get what you can'. It's not just greed, they see it as survival I suppose, but the old fishermen used to say 'if you put nothing back, you get nothing out', and they'd return some offal or something. Now they're trying to take out everything. Occasionally we get herring landed here via the purse netters from Ayr, and odd ones turn up in the Shields fishermen's nets of course. But no-one goes out to catch herring from here now."

Shields' number one fisherman — Cliff Ellis, skipper of the Nielsen. Photo Tony Patterson

PRIDE OF THE FLEET

The Christine Nielsen

Pick up a copy of Fishing News in 1986 and there in the landing totals, number one week in week out for North Shields was the Christine Nielsen, catches above £30,000, a figure so exceeding that of most Shields boats as to put it in a different league. As one crew member said to me; "if it wasn't for the Nielsen, Shields merchants would have been in trouble years ago."

Not everyone would go so far, and by the end of the year skipper Cliff Ellis had stopped having his catch totals in the press — "everyone knowing your business" as he put it. But no-one can argue the fact; he's Shields number one fisherman. Cliff Ellis is 51, occasionally spiky, a smallish restless man, and like most people who achieve anything of note his attitude to work is touched by an obsession. He is dedicated to fishing and at times you sense a sadness, even bitterness that others do not share that dedication. When others haul in and go home, he will persevere. He is fiercely houseproud, keeping his 88ft seine netter spotless; he paces the carpeted floor of a wheelhouse whose combined electronic gadgetry would cost up to £250,000 to replace. The Nielsen wheelhouse is like a video arcade; there is screen after screen of coloured lights, sonar equipment bleeps and pings, radar fingers circle endlessly, the print-out information machine chatters away to itself. You could be on the flight deck of the star ship Enterprise, except the days that were to follow had little to do with the comfortable fantasies of sci-fi. Cliff Ellis spends up to 18 hours a day up here, pacing, checking, more pacing. Most of his life is spent on these few square feet of carpet.

Twelve years ago he helped build this boat which he part owns, as he does a second Shields big boat, the Lindisfarne. My own trip on the Nielsen was approached with a deal of apprehension. It was already mid-November, the North sea was growling like an angry dog, and many boats were hardly bothering to put to sea at all. Not so the Nielsen.

"You could be out for four days, could be fourteen, who's to say?" said the skipper. "And you'll have to find somewhere to sleep. All the bunks are taken." Cliff Ellis's semi-amusement at the rigours I was likely to face and his observations of the resulting traumas became, during the trip, something of an obsession with me, part of the psychological games which long trips can throw up.

The boat waited three days in Shields as the shipping forecasts refused to improve. Impatience grew until finally the skipper said "Bugger 'em. We'll just go. We could wait forever. They're wrong half the time anyway."

And the moment I stepped on board, a gusty Monday morning, things began to go wrong. The Nielsen's a deep draught boat, Shields gut is shallow water, and the engines were no sooner started than the screw (propeller) ensnared itself on something from the bottom. A further 30 minutes frustration waiting for a diver (who would cost £70). Down he went emerging with 12 foot of mooring rope as thick as an elephant's trunk.

We were away, a generous sized leaving party on the quayside waving us off into the dubious delights of an indeterminate journey (time or destination) in weather which threatened to have no sense of humour whatsoever.

"It's a bad time of the year for fishing" said Cliff as he guided his boat out the Tyne, "the sea is already churned up so the fish don't react so much to the disturbance caused by the ropes dragging along the bottom." He has a high-pitched voice, that sense of a nervous excitement I remember from hearing footballer Alan Ball.

Seine netting, or fly fishing, as the Scots call it, is similar to anchor bashing (see the chapter on The Frem), except that instead of working a series of shots round a dropped anchor, the boat traces a triangle round a floating dan, shooting out a mile of rope, then the 260ft net, another mile of rope, and hauling in the two ends which slowly drive the fish into the net. Each shot takes two hours to complete, and there are up to seven shots a day. This is big-time fishing and on this trip I would see sights I'd never seen before. Also, on a trip lasting ten days I'd experience the full psychological impact such deprivation can bring. Fishing, I had long since concluded, was unlike any other occupation, a fact firmly reinforced by my time on the Nielsen.

We steamed East, then South, down towards Dogger Bank, a 17 hour journey with the first shot coming at 5am Tuesday. The Nielsen now records previous trips and shots not on charts but floppy discs, summoned at the press of a button on the video plotter where a small green cursor shows our progress.

She's an impressive vessel, inside and out, a full shelter deck (fitted some years after she was built), which is probably fishing's most merciful recent invention, a galley-cum-messroom with modern electric cooker, fitted kitchen units and two 'fridges, an eight berth cabin below (Cliff Ellis has his own cabin alongside the wheelhouse), a toilet, small sink for washing, and a shower which no-one seems to use. Possibly because preparing for the same in a force eight gale is complicated enough to make it not worth the effort. There's also a ghost called George, which sounds funny but isn't; the aftermath of the Nielsen going to the aid of a tipped over oil rig, the Alexander Keegan which claimed 84 lives. The Nielsen arrived to a scene of harrowing devastation, men frozen to death in life-jackets, some only in underpants. 'George' was the body they pulled on board, a Norwegian clinging to the rig's personnel lift cage. The body lay on deck under a sheet for a day and a half before the navy took him off.

It's a mixed crew; from the 17 year old apprentice Ian Stoneman, nicknamed 'Bugs Bunny' and forced to suffer the usual humiliations and more (see later), up through 22 year old John Burwood, the skipper's son-in-law, 25 year old Shaun Barron, a quiet enigmatic man who spends a deal of time in the fishroom with brother-in-law John Pirie, a gravelly voiced sea dog with tattoes just about rivalling those of 30 year old Keith Burwood, in charge of winding gear whose noise will drive me mad over the next ten days. There's 41 year old John Lota, Portuguese by birth, the bulky waterproofs looking slightly at odds with his fancy cigarette holder. He looks a little like Max Wall, sounds like Chico Marx, he's an ex-skipper and globe-trotter, a rough-edged cosmopolitan whose verbal joustings with Skipper Ellis just about stay on the side of civility.

Then there's Billy Smith, sixty years old (in a job which would kill off many men half that age), and not an ounce of fat on him. He's not only the engineer, he's not only among the fastest gutters on board, he's also the cook/chef, and will serve up, often as if by magic given the elements, as many as 30 meals a day, huge roasts of beef, hash, home-made puddings and pies, delicious soups. On more than one occasion I swear the man's calming influence helped save my sanity. He looks like a windswept Peter Cushing, was once a merchant seaman, took a job on shore, went back to sea as a fisherman at the age of 40, and will probably retire some time if he gets round to it.

Cliff Ellis has a nickname among the crew too. It's Benny Hill.

Fishermen sleep when they can, which sometimes isn't too regular. On the first night's steaming we turn in around 8pm, they to the cabin, me to my two foot wide wooden bench which regularly over the next ten nights will roll me off to the floor. Sleeping at sea is fitful at the best of times, and these aren't the best, especially as each time the night watch changes (about every hour) the new man on watch invades my 'bedroom' to make a cup of tea.

"You're lucky to be here in Winter" says John Lota with a new definition of the word 'luck'. "In Summer we work every moment of daylight, which can mean a 19 hour day, 140 hour week."

I stretch out on my bench, the boat pitches and tosses heading ever further out into the wild North Sea.

And Tuesday morning brings the full realisation of what winter fishing entails for these men. They rise at 5am, silently pulling on boots and waterproofs. First morning and there's a depressed fatalistic air not totally explained by the elements. Outside, without even the first hint of dawn, there is sleeting driving rain. The sea is a black roaring monster, smacking its foam-mad lips. Twenty foot high walls of spray rear up at the dipping prow and hiss onto the wheelhouse window. They intend to fish in *this* . . .?

Yes. The first rope is shot from one of the three giant winches under the shelter deck, winding out half the length of the boat, and watch you don't walk into it. The net, stacked on the transom stern (the raised rear open section of the boat) is then attached and itself shot, the second rope sent out, and the boat moves in to collect the dan (buoy) which was dropped at the start of the operation. The long process of winding in begins, and with it that high endless drone of the winding gear which drills its way increasingly into your head, neatly accompanied by the steady thump-thump of the engine.

The morning routine is always the same. Shoot the nets, eat breakfast, snatch a few moments of sleep if possible (it may be the only chance for 17 hours), then haul in. Except even the best routines go wrong and half-way through Bill Smith's tasty fry-up, there's a shout from the wheelhouse. The net is caught fast, probably on the merlog, a seabed forest which can ensnare and snap your gear. Half-eaten breakfasts are abandoned as the crew hurry on deck. I sit and stare round the table at this scene resembling something from the Marie Celeste, the crockery, the cutlery, the food, but no people. By the time the net is freed, the breakfasts are cold.

Shooting nets is a delicate job, especially in these seas. Up in the wheelhouse (which often seems remote from the deck, connected mainly through an intercom over which Cliff Ellis's voice comes as a piercing shriek), the skipper plots each tiny movement. Keith Burwood watches the ropes, ringing a large brass bell to indicate each required change in direction. The ropes' position is vital. If they don't haul in at exactly the right tension and angle the net turns and releases its haul. This will happen on our very last shot of the journey, a bountiful hunting ground, and we'll lose up to £2,000 worth of fish.

Among the Nielsen's wonders is the power block, a monstrous orange crane affair which hangs over the stern like some giant creature's claw. Over its hydraulically driven bobbins the net is slung and hauled in. This strange appendage imbues the activity with a sense of ritual. The power block moves up, down, sideways, controlled by Bill Smith, and its movements make it yet more alive, animated. It dwarfs the men below who seem to be gathered in primitive worship at its feet. When the ropes are winched in, and the net's four lines slung over the bobbin, the sight of four fishermen in gleaming waterproofs, to the background of a large swelling sea, tugging like bell-ringers below the great nodding head of the crane evokes some sort of pagan ceremony. The bent claw swings from the stern to the starboard side as the cod end rises into view on the sea's surface. Now it's winched up on the gilson after the fish are shaken down the length of the net. The cod end swings dangerously like some silver demolition ball. Catch a swipe from it at this stage, and you'd be overboard.

During one lift it knocks poor Ian clean off his feet, but luckily into the boat. There can be a ton or more of fish swinging away there, and it's someone's job to duck underneath, release the cod end knot and allow the fish to spill out. Any volunteers?

The first day's hauls are poor. The fish slither and slide into the pound where, without ceremony, they're shovelled or kicked or thrown towards the shelter deck, and temporarily boxed before gutting. Again, it's mainly codling, but some lemon sole, witch, mackerel, monk, catfish. There's one species, the gurnard, a spiked alarming looking fish who is, as they say not wanted on board. Hundreds of "gurneys" come aboard each catch. They all die. They're all chucked back. More waste.

If the shelter deck is a blessing, it's a mixed one. The metallic cover keeps off the sea, but also acts as an echo chamber for the roar of the winding gear and the constant throb of the engines. It also traps the foul odours that are part and parcel of gutting. There's the sickly smell of the fish intestines, plus the nauseous smell of diesel oil (surely one of the most unpleasant nasal memories of fishing). On this first day, with small hauls, the shelter deck isn't seen in its full glory. Later, when we took hauls of up to 60 boxes and production moved into top gear, the experience is a sobering one. In a deafening din, the floor slanting this way and that as the heavy seas buffet us, several thousand fish are gutted at lightning speed. The sea water running in and out through the scuppers (small holes in the boat's side) plus the regular jet from the 'donkey' (the boat's hose) send a disgusting wave of blood and guts sloshing about ankle high from one side of the boat to the other. Look out through a hatch at the escaping effluence and it resembles something spat from the mouth after a dentist's extraction. It is hard fully to convey this sensation. Decked out in waterproofs, you rapidly become covered in slime and blood. After gutting (and mine is considerably slower than theirs, a fact they never once ribbed me about) each fish is tossed high into the fish-washer, a piece of machinery I'd not seen on previous trips. A high pressure jet of water swirls round an open drum at the top. The fish are tossed in here and after a few 'circuits' slide down a ski-slope into the fish room below where John Pirie and John Burwood wait to ice and box them. It's a strange sight on this shelter deck, six men tossing bloody and savaged fish high into the air, as if at some bizarre fairground stall. While you are gutting the net is being shot again, and while one rope runs the length of the boat, the other is shot through a starboard hatch. In these heavy seas it often acts as a sneaky entrance for giant waves which can thud against you like a Frank Bruno straight left.

Gutting is probably the least savoury part of a fisherman's work. Often the fish is still alive, quivering as you push the knife in under the gill and slit down the length of the belly.

Pushing home the knife into still living flesh, not once, not twice, but time after time is initially horrendous, slowly less gruesome, and, in the way that human beings adapt to all manner of atrocities, in time just a matter of course. What at first makes me recoil with moral disgust, I eventually accept as 'part of the job', and realise how much our reactions depend on circumstances and conditioning.

The fish often jerk and quiver on throughout the gutting. You can sometimes see them quiver as they swirl round the washer, and strangest of all, down in the fish-room, where they are iced, boxed and stacked in rows, I have seen them, beneath the ice, beneath the sheets of greased paper, and surely long dead, I have seen them twitch more. It is akin to seeing the corpses jerk in the morgue.

But this is a bad day's fishing, the sea too wild to offer up its bounty. There is one moment of oddity on this day, an incident both comforting yet disconcerting. The nets are shot and hauled in fierce winds, in high rolling seas. Standing at the stern, water running from my waterproofs, eyes stinging from the spray, I stagger along the boat's side, and make my way into the galley. The boat is tossing and heaving, keeping your feet is difficult, the decks are awash, and there, inside the galley, as calm as a Tibetan monk, rolling out pastry for the apple pie he'll serve up that evening is Bill Smith. The combination seems extraordinary. I stand and watch him crimp the edges, rub lightly with milk, I watch him sprinkle on sugar. He places the pie carefully in the oven while beyond that door the sea and the winches roar. Apple pie for tea.

Fishermen *are* the last of the hunters, and after two poor hauls, that is exactly what we do — we go hunting. By now we're 140 miles from the Tyne in Dogger Bank. All the electronic gear in the world won't help when the fish aren't there. I am aghast that men are out here, that they are fishing in such conditions. To them the elements are a matter of course. Cliff Ellis paces the floor of the wheelhouse. Like most fishermen he's expert at walking on a rapidly shifting surface, keeping your balance on something akin to a fairground cakewake (on most trips I am covered in bruises by day two). He leans and adjusts to the swell, an automatic action. At times the back of his head seems almost to be brushing the floor such is the lean. "Bad hauls" he says. "The worst day's fishing this year" says John Pirie. We move ground and at 5pm suddenly the fish are there. The cod end comes up hugely swollen, the fish spill out into the pound and we're all wading knee-deep in them.

On this first day, gutting is still traumatic enough almost to make me vomit, the scenes of bloody carnage. Machines can do it, but fishermen are suspicious of them, claim they're unreliable (innovations are absorbed slowly in this world). It's not only the carnage (if we can use such a word about fish), it's also the waste.

Almost 50 per cent of each haul is unsuitable, wrong fish, wrong size. Hundreds upon hundreds of dead fish are kicked and hosed back into the sea, young codling which should be tomorrow's breeders, undersized haddock, whiting. This is just one boat on one day, yet I see thousands of uselessly dead fish. Multiply this a thousand fold, imagine the effect on the stocks. Cliff Ellis agrees, yet points to the fact that his mesh size is already 10-15cms larger than the legal requirement.

"So what am I supposed to do?"

To see this futile waste is to realise that someone, somehow must take drastic action, legislation to prevent the mass devastation of the seas. Not only is such wastage economic madness in the long-term, not only is it morally wrong to kill so many creatures needlessly, it is also hugely inefficient on the boat, much of the time spent sorting and discarding the unwanted fish. I wonder if those who make the laws ever witness such waste at first hand? Have they ever seen the regular mountains of dead discarded fish which should shame a world of such starvation? And has a fishing minister ever *been* on a fishing boat?

When the long day is done, which on Tuesday is around 10pm, the work-place is still there. You cannot leave it behind, you are still being tossed around in the same small craft in the same hostile environment. No place to go, no other faces to see. Even miners can wash off their pit black and go down the pub.

Cliff Ellis cuts a lonely figure in the wheelhouse. He's been at sea for 35 of his 51 years, came to Shields from Grimsby 26 years ago because as he says, "I liked the look of the place, I thought I could make things happen in Shields, and I thought I could make money."

"I'd worked several years for virtually nothing, and I'd done stints of up to 32 days anchor bashing, but I'm dedicated to this life. Oh, it's no life at all in many ways, our lives are wasted out here at sea, but I'm dedicated. This boat is my home for three quarters of the year, so yes I make sure everyone looks after it."

He's skippered other boats, The Contessa, The Conmoran, The Concorde, The Contessa a second time. Crew members such as Bob Shaw, as affable a fisherman as you'd come across, have followed him from boat to boat, and earnings on the Nielsen — probably three times the earnings of many Shields fishermen — means there's never a shortage of willing deckies. This is the pride of the fleet.

"Yet I see what's happening. We're catching codling now because the cod have mainly gone. Then we'll have to go onto haddock when the codling go. Where then? Yes, it's our own fault, we've overfished the seas, and now we're being made to pay."

"But increase the mesh size too rapidly and you immediately put a lot of small fishermen out of business" and again he asks the question: "so what do you do?"

In 1982 the Nielsen almost sank, a collision off the German island of Heligoland (a duty-free paradise for visiting fishermen). She was hit by a container vessel in thick fog, the container bearing down on them at a suicidal 17 knots ripping into the port side, chopping off the stem. There was pandemonium, smashed glass, crew thrown everywhere. The old sea charts still bear the blood marks of crewman Johnnie Hook (sic) injured in the collision. Cliff Ellis still has a hole in his leg.

They limped home to Shields under escort after initial fears they would sink.

"I suddenly saw this boat loom out the fog and knew it would hit us. I brought the boat astern to lessen the impact. If I hadn't done that, and it had hit midships, we'd all have been finished."

Breakfast on the Nielsen is early, the main meal somewhere between two and three pm depending on the hauls. Given time Billy Smith often serves up snacks too. Eating habits here vary. John Pirie and John Burwood reduce all meals to a sandwich, meat, veg, even roast potatoes picked up and squeezed between thick slices of white bread.

Another sight for sore eyes in the messroom; the same Billy Smith patiently rinsing through the gingham table-cloths. Washing the dishes is traditionally the apprentice's job. Now how an apprentice ever sticks it out to become a full fisherman I shall never know. They are asked to suffer the insufferable. Initiation entails unspeakable things being done to them with treacle (need I expand?). On this journey poor Ian is hounded constantly. His food is hidden, his hat is pinched, his trousers are yanked down, he is verbally insulted, physically knocked around, his masculinity and preferences are questioned loud and long. He accepts this with a sort of glassy-eyed tolerance; "it's something everyone goes through, I'm not really any different."

Fishermen and writers are not usual workmates,, and however good the intentions, I often sense an alienation. An incident happens on this trip sparked by a casual remark which creates a bond I had not previously experienced and which fortified me through the long days and nights.

"Poet and all are you?" asks Keith Burwood on day three; "so why not write a poem about us?" The idea takes root, and that night, the sea a dark beast roaring to be let in, and the crew asleep down below I pen an eleven stanza comic ballad about the Nielsen (see end of chapter). There's a stanza about everyone, including the skipper, and the response when I read it aloud at breakfast the next morning delights me. Suddenly my writing is not merely remote, but to do with them. I've seen them at work. Now they've seen me.

Things improve.

Fishing is a filthy business, but the Nielsen is the cleanest, best maintained boat I've been on. After each haul, when the blood splatters the walls, the shelter deck roof, the lights, when guts and innards are everywhere, the entire work area is hosed down from top to bottom until not the merest red streak remains. They hose themselves down too, to stand dripping like saturated sponges, but clean.

At times this operation is only minutes before the next haul, such is the length of gutting and boxing, and rapidly the scene once again resembles a slaughterhouse. No matter. The process is repeated again and again, every vestige of gut, every entrail, every squashed fish head removed. The work rate here is phenominal. Gutting, washing, boxing — how many other areas of British industry could claim such speed and efficiency. Within one hour many thousand fish are taken from the sea, gutted, cleaned, sorted into sizes (there are six cod sizes alone) sorted into types (there's the occasional 'loner' such as halibut or turbet), iced, boxed and stored for the journey. Fishermen work when there is work to be done, meal breaks, tea breaks, stoppages, overtime, these are non-existent concepts out here when there's work.

Later that day we move North from Dogger into the Argyle oil fields. The fishing improves. The first day 48 boxes, the second 81, 90 on the third, and 130 the fourth. By day four there is only the boat, the sea, the fish, and each other. Fish are strange creatures, we haul up a turbet, huge milky white face and a formless black body like an ink blot. And look at dog fish, not soft and slimy like most fish, but skin so rough men once used it as sandpaper; and a fierce sharp spike half hidden near the rear fin, a spike which draws blood on this trip from at least one crew member.

There is much grinding repetition, shoot, haul, gut, shoot, haul, gut, a routine acted out against a sea now given over almost constantly to gale force winds. The shipping forecasts have an ominous predictability; "gale warnings in all areas . . ." and the kind of seas which a short while ago I would have viewed with absolute panic I now view almost stoically, and only occasionally does that panic well up inside, the sudden knowledge of how precarious and isolated we are. That same night as the boat is riding and crashing, threatening to shake itself to bits on this wild sea we hear May Day messages for a boat stricken off the Irish coast. That same night a boat sinks North of us in the North Sea, and on the next night an oil platform slips sickeningly on an 11 degree lurch.

Engine problems; a broken fuel injector, and a bar supporting the cooling pipes has cracked, there's a broken drum chain. In the safe confines of a dry dock, none is a major repair. Tackling them in a force eight isn't quite so simple. John Lota and Bill Smith fix it.

"It looks" smiles skipper Ellis "like being a long trip. Wish you hadn't come?" I stare gloomily at the horizon gashed by waves, wonder if there's any life beyond it.

It's one of our gruesome little games; Cliff Ellis pokes at my obvious vulnerability to this life. I dig in my heels.

On this fourth day, we begin at 5.30am, the final fish are gutted and boxed at 11.15pm. By mid-evening when we are all done in, when the day seemed finally and painfully to be drawing to a close, Cliff Ellis says "another shot" and we shoot again, in the dark. The crew finally crawl to their bunks, to be woken again at 5am. I am sure Cliff Ellis sleeps sometimes; it's just that I'm never aware of it, only the restless pacing, pacing of the wheelhouse, the constant studying of his screens and sonars.

And look at fishermen's hands. Bob Hall, who takes a protective interest in this landlubber throughout the trip, holds them out for me to examine. Years of gutting have taken their toll. He can no longer get on his wedding ring. The fingers are gnarled and twisted like dead tree trunks. Many of his fingers will never be straight again. Salt water boils which are irritated by material chafing on skin throw up angry red rashes as the salt worms its way down to the roots of the tiny hairs.

But there's excitement too, on day five as we take a monster haul. The cod end is so full that when it's pulled up on the gilson, the boat leans crazily from the weight. The haul fills 100 boxes on deck (this number can be almost halved by the time they're gutted and sorted) and with this quantity, the work takes on an even more frenetic quality. Even as they are being sorted Keith Burwood is running out the winding gear, the next shot is underway. No matter how big the haul, never once do I see them defeated. Always the last fish is cleared before the new haul comes aboard. And always the deafening din, almost every word you exchange has to be repeated, an increasingly wearisome practice.

To me one stretch of stormy sea is like another. A good skipper knows his charts, his tow lines, his fish marks. He knows when to shoot, when not to,

"Pride of the Fleet" — the Christine Neilsen docked at North Shields Quay.

Photo Tony Patterson

knows when to stay, when to move on. At a time when many fishermen are safe in port, not even contemplating a trip, here is the Nielsen, atop of a wild sea, and taking huge hauls. This boat once returned to Shields with 1,250 boxes, 450 on the final haul. The huge fishroom was jammed tight, and 280 boxes stored on deck. It's the kind of tale any fisherman loves to tell.

Quirky moments too. Cliff Ellis indicates our position on a chart, an area called The Patch. It's only seven fathoms deep, here in the middle of the North Sea, maybe twice the height of an Olympic diving pool. "If we sink now" he jokes "we just climb to the top of the mast and wait to be picked up."

During the great sprats bonanza, Ellis was the Sprats King, the first skipper to take on board the new technology, which saw him returning to Shields with as much as 70 or 80 tons of sprats a day.

"A lot of people in Shields don't like me" he says. "It's up to them. I just get on with my work" and he emerges only occasionally from the wheelhouse, a quick snack, a few words in the messroom, then back to his wheelhouse, his instruments, back to his hunting.

With the exception of the apprentice, who is on a fixed rate, the crew are all share fishermen. For every £1,000 earned on the boat, each crew member will get about £49. They have no union — why not?

"I don't think most fishermen want one" says John Pirie. "They tried it once, it didn't work, bloody union stopping a boat from sailing because of some technical detail. It's between you and your skipper, you're dependent on one another out here, no shorebound bloke can sort that out with you. That's what being a fisherman is, if you don't like it, don't do it. You're a gambler and you gamble with your life and with your money. We don't need interference from outsiders."

When he was 16, John Pirie's brother was washed overboard and drowned from a Grimsby trawler. John was 18. Despite their appalling conditions, I found few fishermen eager for a union. Attempts to form their own, plus attempts to embrace them into the Transport and General, have failed.

On day seven, I believed I was going mad. This was now the longest trip and psychologically the most testing. In the early hours, I awoke from my narrow bench to the chug of the engines, the thud of the sea against the side. Each morning I would awake stiff and aching, and grab a couple of hours in a bunk during the first shot, relief in many ways, though the proximity of the winding gear to the cabin makes it a deafening place. I was beginning to appreciate the nature of deprivation, the removal from life of those factors which make it more than mere existence. I had been on the boat an eternity. Another eternity seemd to stretch up ahead. Friends, home, pleasures, I viewed them all as if through the wrong end of a telescope. No escape, no privacy, not for a single moment in 24 hours could you flee this claustrophobic and miniscule world that was your entirety. You awake exhausted from a fitful sleep at the sinking realisation that another long day of the same grinding repetition awaits. And for a moment, for no fault of their own, the crew seemed people I could no longer face. The incessance of their presence, nine people I had never previously met, and with whom I had only limited things in common, this forced, captive kind of companionship seemed intolerable.

Shooting, hauling, gutting, the ceaseless drone of the engines and winding gear, the blood, the smell, the ground constantly moving beneath your feet, the inability to break free of these confines. What sort of world was this? How did fishermen possibly survive it?

"Miss home?" A crew member read my mind. He laughed "That's why I come out here, to get away from it" and still I don't know how much truth there was in that statement. Home would have to be pretty bad.

Cliff Ellis pats me on the back. "I once had this television crew come out, they planned a 48 hour trip. Within hours of leaving the Tyne their equipment was all over the place. They were, and this is true, on their hands and knees imploring me to take them back."

"I took pity on them, took them back."

On the eighth day, we almost lose four of the crew. One sudden moment. The first shot of the day, dawn still struggling through, they're standing on that shallow transom stern, pulling the lines over the power-block bobbin, when a giant wave comes rolling and growing from the port side. It takes a big breath then explodes itself at the four 'bell-ringers', picks them up and hurls them across the boat. John Burwood is thrown half-way up the power-block and left with a dead leg. Bob Hall clings desperately to the gunnels. The stern is a giant bowl of water. Had the wave been at a different angle it would have thrown them not into, but out of the boat, when they would surely have been finished, for to rescue anyone from this violent seascape of sudden mountains and valleys would be impossible. All four are drenched, the wave has run through the aft door and now sloshes about in the hallway. The transom stern has always seemed a precarious place to work, a place where often men are left with nothing to hold or secure them. Why do they have no lifelines attached to them (a growing habit in some countries)? Why do they wear no life-jackets?

"They would get in the way" says John Lota and shrugs his shoulders. They all agree. They are all mad.

In the churning sea, a shoal of porpoises knits its way past the boat. And in the wheelhouse, a micro-chip disaster. The computer has 'eaten' the shot records for the entire year. They will all have to be put on tape again from the old-fashioned charts, which nothing ever eats. Day eight and everyone looks sick of the trip. We all want to go home.

"Alright, Wednesday morning we go in" says Cliff, and at last we have a definite time, two days time, forty eight hours . . . Bill Smith fingers his brass key ring. Its legend reads "Old Fishermen Never Die — They Just Smell That Way". He pops a side of beef into the oven. At the end of day eight the crew wriggle out of their waterproofs (not the easiest of tasks). We all feel low.

"This used to be a sport" says Cliff Ellis. "Now it's too competitive. Once I'd move 100 miles looking for fish, but I can't afford it now, I have to plan each shot meticulously." Even so, on this trip we will have journeyed 600 miles, into Fisher, Humber, Dogger, as far East as German Bight, we shall have taken on board 520 boxes of fish in conditions which have been hostile virtually throughout. It's a true fishing boat, and these are true fishermen.

Come the Tuesday dawn, the last one we shall see, and there's a magnificent sunrise, the entire eastern sky stained such an indelible red as to infuse the sea itself with its reflected colour and turn it into a churning mass of blood.

"A wasted life" muses Cliff Ellis, staring from the wheelhouse. "What else can it be, spending three quarters of it out here?"

Now something extraordinary happens, our final chapter. The last day sees big hauls, we work through until 6pm, then set course for the Tyne at eight knots, an eight hour journey. Our fishing routes have slowly taken us nearer home for the last two days. and it's cleaning time.

The Nielsen never goes back to port without every inch of her being scrubbed, polished, cleaned, wiped. How bizarre, out here to find the smell of guts and diesel temporarily replaced by that of lavender polish. In the wheelhouse the sight is of the region's no. 1 skipper on his hands and knees like Mrs Mopp, scrubbing the carpet with hot soapy water. For more than three hours every crew member sets to. Every inch of brass is polished, every wall is cleaned, every floor, every stair, every surface, every inch of glass is washed, the entire boat is spring-cleaned. The crew confess to hating this concentrated domesticity, and it's an excess which maybe symbolises the dedication of the skipper, but the end product is a brilliantly gleaming vessel, a proud craft to return to Shields. You'll not find a fingerprint nor dirty smudge, in the cabin, the messroom, the wheelhouse.

It's been a long trip, often horrendous. In my bag I have such strange souvenirs as a piece of transparent plastic, almost like a ruler. Who would guess this to be a squid's backbone (thanks, John Burwood). Bob Hall cut open a sea urchin and presented me with a cluster of tiny penguins — they're almost exact replicas of the dinner-suit creature. We land at 3am, ironically the final miles in a calm sea. For the crew the journey hasn't ended. There's another 12 hours at the quay, landing the fish, sorting boxes, checking nets, and taking on ice for the next journey. This is Wednesday, and they have the unashamed luxury of two nights ashore. Then it's back to that sea.

The Ballad of the Christine Nielsen

It was on the good ship Nielsen
and the crew was feeling sore
they'd been at sea for three days
and the boxes numbered four.

Skipper Ellis paced the wheelhouse
trying to understand the scene;
he'd got such lovely colours
on every video screen!

Bill Smith slumped in the galley
needing some relief
he'd started going crazy
and gutted a side of beef.

John Burwood scratched his blond hair
"no bloody fish!" he said
biting off his fingers
between two slices of bread.

And look at poor John Lota
this worry makes him older,
sucking away for hours
with no fag in his holder.

In the fish room there's John Pirie;
he's got only ice to lug
there's more bad news for Johnnie
someone's drinking from his mug.

His dead-eyed mate Shaun Barron
clings to just one futile wish
that the silly bloke called Mortimer
would stop gutting tiny fish.

"Cheer up!" cries Bob Hall, laughing
"I've brought us some great treats!"
He winks at all and sundry
producing three boiled sweets.

In his bunk young Ian Stoneman
has had all that he can take.
He says the boat's a nightmare;
it's lucky he's never awake.

Standing by the winches
Keith Burwood's given up hope
wondering why, with no fish
he's reeling out miles of rope.

The Nielsen sails the high seas
the crew slumps off to bed
dreaming of a huge catch
waking up to sandwich spread.

Peter Mortimer

The Last Shot . . . a crowded North Shields fish quay — the view down from Dockwray Square. *Photograph Tony Patterson*